Marx Joyce
Hardy Emerson Austen
Defoe Abbott Machiavelli Chesterton Cooper Hugo
Melville Montaigne Haggard Eliot Grimm
Stoker Christie Molière
Wilde Carroll Byron Schiller
Garnett Maupassant Engels Smith Kafka
Goethe Fitzgerald Hawthorne Hall
Cotton Einstein Dostoyevsky Willis
Baum Henry Kipling Doyle
Leslie Dumas Nietzsche Balzac
Flaubert Turgenev Crane
Stockton Vatsyayana Verne
Burroughs Gogol Vinci
Curtis Tocqueville Whitman Busch
Homer Widger Tolstoy
Darwin Thoreau Twain Plato Scott
Potter Freud Zola Lawrence Harte
Kant Jowett Stevenson Dickens Hesse
Andersen Burton
London Descartes Cervantes Voltaire
Poe Aristotle Wells Cooke
Hale James Hastings
Bunner Shakespeare Irving
Richter Chekhov Chambers
Doré da Shaw Benedict Alcott
Dante Wodehouse Pushkin
Swift Newton

One Thousand Questions in California Agriculture Answered

Edward James Wickson

Imprint

This book is part of TREDITION CLASSICS

Author: Edward James Wickson
Cover design: Buchgut, Berlin – Germany

Publisher: tredition GmbH, Hamburg - Germany
ISBN: 978-3-8424-2795-2

www.tredition.com
www.tredition.de

One Thousand Questions in California Agriculture Answered

By E. J. Wickson

Professor of Horticulture, University of California; Editor of Pacific
Rural Press; Author of "California Fruits and How to Grow Them"
and
"California Vegetables in Garden and Field," etc.

Foreword

This brochure is not a systematic treatise in catechetical form intended to cover what the writer holds to be most important to know about California agricultural practices. It is simply a classified arrangement of a thousand or more questions which have been actually asked, and to which answers have been undertaken through the columns of the Pacific Rural Press, a weekly journal of agriculture published in San Francisco. Whatever value is claimed for the work is based upon the assumption that information, which about seven hundred people have actually asked for, would be also interesting and helpful to thousands of other people. If you do not find in this compilation what you desire to know, submit your question to the Pacific Rural Press, San Francisco, in the columns of which answers to agricultural questions are weekly set forth at the rate of five hundred or more each year.

This publication is therefore intended to answer a thousand questions for you and to encourage you to ask a thousand more.

E. J. Wickson.

Part I. Fruit Growing

Depth of Soil for Fruit.

Would four feet of good loose soil be enough for lemons?

Four feet of good soil, providing the underlying strata are not charged with alkali, would give you a good growth of lemon trees if moisture was regularly present in about the right quantity, neither too much nor too little, and the temperature conditions were favorable to the success of this tree, which will not stand as much frost as the orange.

Temperatures for Citrus Fruits.

What is the lowest temperature at which grapefruit and lemons will succeed?

The grapefruit tree is about as hardy as the orange; the lemon is much more tender. The fruit of citrus trees will be injured by temperature at the ordinary freezing point if continued for some little time, and the tree itself is likely to be injured by a temperature of 25 or 27° if continued for a few hours. The matter of duration of a low temperature is perhaps quite as important as the degree which is actually reached by the thermometer. The condition of the tree as to being dormant or active also affects injury by freezing temperatures.

Under certain conditions an orange tree may survive a temperature of 15° Fahrenheit.

Roots for Fruit Trees.

I wish to bud from certain trees that nurseries probably do not carry, as they came from a seedling. Is there more than one variety of myrobalan used, and if so, is one as good as another? If I take sprouts that come up where the roots have been cut, will they make good trees? I have tried a few, now three years old, and the trees are doing nicely so far, but the roots sprout up where cut. I am informed that if I can raise them from slips they will not sprout up from the root. Will apricots and peaches grafted or budded on myrobalan produce fruit as large as they will if grafted on their own stock?

Experience seems to be clear that from sprouts you will get sprouts. We prefer rooted cuttings to sprouts, but even these are abandoned for seedling roots of the common deciduous fruits and of citrus fruits also. The apricot does well enough on the myrobalan if the soil needs that root; they are usually larger on the peach root or on apricot seedlings. The peach is no longer worked on the myrobalan in this State. One seedling of the cherry plum is about as good a myrobalan as another.

What Will the Sucker Be?

I have a Japanese plum tree which bears choice plums. Three years ago a strong young shoot came up from the root of it, which I dug out and planted. Will it make a bearing tree in time and be of like quality with the parent?

It will certainly bear something when it gets ready. Whether it will be like the parent tree depends upon the wood from which the sucker broke out. If the young tree was budded very low, or if it was planted low, or if the ground has been shifted so as to bring the wood above the bud in a place to root a sucker, the fruit will be that of the parent tree. If the shoot came from the root below the bud, you will get a duplication of whatever stock the plum was budded on in the nursery. It might be a peach or an almond or a cherry plum. Of course you can study the foliage and wood growth of the sucker, and thus get an idea of what you may expect.

Tree Planting on Coast Sands.

I wish to plant fruit trees on a sandy mesa well protected from winds about a mile from the coast. The soil is a light sandy loam. I intend to dig the holes for the trees this fall, each hole the shape of an inverted cone, about 4 feet deep and 5 feet across, and put a half-load of rotten stable manure in each hole this fall. The winter's rains would wash a large amount of plant food from this manure into the ground. In March I propose to plant the trees, shoveling the surrounding soil on top of the manure and giving a copious watering to ensure the compact settling of the soil about and below the roots. The roots would be about a foot above the manure.

On such a light sandy soil you can use stable manure more safely than you could elsewhere, providing you have water handy to use if you should happen to get too much coarse matter under the tree, which would cause drying out of the soil. If you do get plenty of water to guard against this danger, you are likely to use too much and cause the trees to grow too fast. Be very sure the manure is well rotted and use one load to ten holes instead of two. Whether you kill the trees or cause them to grow aright depends upon how you use water after planting.

A Wrong Idea of Inter-Planting.

What forage plant can I grow in a newly planted orchard? The soil is on a gently inclined hillside - red, decomposed rock, very deep, mellow, fluffy, and light, and deep down is clayish in character. It cannot be irrigated, therefore I wish to put out a drought-resisting plant which could be harvested, say, in June or July, or even later. I find the following plants, but I cannot decide which one is the best: Yellow soja bean, speltz, Egyptian corn, Jerusalem corn, yellow Milo maize, or one of the millets. What do you think?

Do not think for a moment about planting any such plant between orchard trees which are to subsist on rainfall without irrigation. Your trees will have difficulty enough in making satisfactory growth on rainfall, and would be prevented from doing so if they had to divide the soil moisture with crops planted between them. The light, deep soils which you mention, resulting from decomposed rock, are not retentive enough, and, even with the large rainfall of your region, may require irrigation to carry trees through the latter summer and early fall growth.

What Slopes for Fruit?

I want to plant some apples and berries. One man says plant them on the east or south slope of the hill and they will be ripe early. Another man says not to do that, for when the sun hits the trees or vines in the morning before the frost is off, it will kill all the blossoms, and as they would be on the warm side of the hill they would blossom earlier and there will be more frosts to injure them. I am told to plant them on the north or west side of the hill, where it is cold, and they will blossom later and will therefore have less frosts to bother them, and the frost will be almost off before the sun hits them in the morning.

Fruit is grown on all slopes in our foothills, depending on local conditions. On the whole, we should choose the east and north slopes rather than the east and south, because there is less danger of injury from too great heat. In some cases what is said to you about the less danger of injury from frosts on the north and west slopes would be true. All these things depend upon local conditions, because there is so much difference in heat and frost and similar slopes at different elevations and exposures. There can never be a general rule for it in a State so endowed with varying conditions as California is.

Trees Over Underflow.

I have planted fruit trees near the creek, where they do not have to be irrigated as the ground there holds sufficient moisture for them, but a neighbor tells me that on account of the moisture being so near the surface the trees will not bear fruit well, although they will grow and have all the appearances of health.

Shallow soil above standing water is not good for fruit trees. A shallow soil over moving water or underflow, such as you might expect from a creek bank, is better. The effect of water near the surface depends also upon the character of the soil, being far more dangerous in the case of a heavy clay soil than in the case of a light loam, through which water moves more readily and does not rise so far or so rapidly by capillary action. If the trees are thrifty they will bear when they attain a sufficient age and stop the riotous growth which is characteristic of young trees with abundant moisture. If trees have too much water for their health, it will be manifested by the rotting of their roots, the dying of their branches, the cropping out of mushroom fungi at the base and other manifestations of distress. So long as the tree is growing well, maintains good foliage to the tip of the branches and is otherwise apparently strong, it may be expected to bear fruit in due time.

The "June Drop."

I am sending four peaches which are falling off the trees. Can you tell me how to prevent falling of the fruit next year and what causes it?

It is impossible to tell from the peaches which you send what caused their falling. Where fruit passes the pollination stage successfully, as these fruits have, the dropping is generally attributed to some conditions affecting the growth of the tree, which never have been fully determined. It is of such frequent occurrence that it is called the June drop, and it usually takes place in May in California. As the cause is not understood no rational preventive has been reached. A general treatment which consists in keeping the trees in good growing condition late enough during the previous season, that is, by seeing to it that they do not suffer from lack of moisture which causes them to close their growing season too soon before preparation for the following year's crop is made, is probably the best way to strengthen the tree for its burden.

Trees Over a Gravel Streak.

I have an apricot orchard seven years old. Most of the land is a fairly heavy clay with a strip of gravel in the middle running nearly north and south. The trees on the clay bear good crops, but those on the gravel are usually much lighter in bearing and this year had a very light crop. Can you tell me of anything I can do to make them bear? The trees are large and healthy looking, and grow big crops of brush.

We should try some water in July on the gravel streak, hoping to continue activity in the tree later to induce formation of strong fruit for the following year. On the clay loam the soil does this by its superior retentiveness.

Fruit and Overflow.

I have 16 acres of rich bottom-land that overflows and is under water from 24 to 48 hours. I would like to set the ground to fruit trees, either prunes, pears, apricots, or peaches. Would it be safe to set them on such land?

Fruit trees will endure overflowing, providing the water does not exclude the air too long and providing the soil is free enough so that the soil does not remain full of water after the surface flow disappears. If the soil does not naturally drain itself and the water is forced to escape by surface evaporation, probably the situation is not satisfactory for any kind of fruit trees. Overflow is more likely to be dangerous to fruit trees during the growing season than during the dormant season, and yet on well-drained soil even a small overflow may not be injurious on a free soil, if not continued too long. Prunes on plum root, and pears will endure wet soil better than apricots or peaches.

Fruit Trees and Sunburn.

How long is it wise to leave protection around young fruit trees set out in March in this hot valley? The trees are doing well, but we could not tell when to take away protection.

It is necessary to maintain the protection from sunburn all through the autumn, for the autumn sun is often very hot, and as the sap flow lessens, the danger of burning is apparently greater. The bark also must be protected against the spring sunshine, even before the leaves appear. So long as the sun has a chance at the bark, you must protect it from sunburn.

Replanting in Orchard.

Is it considered a good plan to set the tree at once in the place where one has died, or is it better to wait a year before replacing?

It is not necessary to wait a year in making a replanting. Get out all the old roots you can by digging a large hole, fill in with fresh soil, and your tree will accept the situation.

Whole Roots or Piece Roots.

For commercial apple orchards which is preferable, trees grafted on piece roots or on whole roots? On behalf of the piece-root trees it is claimed they sprout up less around the tree. On the other hand, it is claimed they never make a vigorous tree. What is the truth?

Value depends rather upon what sort of a growth the tree makes afterward than upon what it starts upon. Theoretically perhaps a whole-root tree may be demonstrated to be better; practically, we cannot see that it becomes so necessarily, because we have trees planted at a time when the root graft on a piece was the general rule in propagation. After all, is it not more important to have soil conditions and culture of such character that a great root can grow in the orchard than to have a whole nursery concentrated in the root of the yearling tree? As for the claim that a root graft on a piece-root never makes a vigorous tree, we know that is nonsense.

Planting Deciduous Fruit Trees.

In order to gain time, I have thought of planting apples and pears this fall, in the belief I would be just that much nearer a crop, than though I waited until next spring. The land is sandy loam; no irrigation. Would you advise fall or spring planting? If fall, would it be best to plow the land now, turning in the stubble from hay crop, or wait until time to plant before plowing?

You will not be any nearer a crop, for next summer's growth will be the first in either case. On land not liable to be too wet in winter, it is, however, best to plant early, say during the month of December, if the ground is in good condition and sufficiently moist. If the year's rainfall has been scant, wait until the land is well wet down, for it is never desirable to plant when the soil is not in the right condition, no matter what the calendar may say. On a sandy loam early planting is nearly always safe and desirable. On lands which are too wet and liable to be rendered very cold by the heavy January rains, planting had better be deferred until February, or as soon as the ground gets in good condition after these heavy rains. Whenever you plant, it will be desirable to plow the land either in advance of the rains, if it is workable, or as soon as rain enough comes to make it break up well. It is very seldom desirable to postpone plowing until the actual time of planting comes.

Budding Fruit Trees.

Is it better to bud in old bark of an old tree or in younger wood bark?
How do you separate old bark without breaking it in lifting the bark?

Buds may be placed in old bark of fruit trees to a certain extent. The orange and the olive work better that way than do the deciduous trees, although buds in old bark of the peach have done well. They should, however, be inserted early in the season while the sap flow is active and the old bark capable of lifting; if the bark sticks, do not try budding. In spite of these facts, nearly all budding of deciduous trees is done in bark of the current year's growth.

Starting Fruit Trees from Seed.

How shall I start, and when, the following seeds: Peach, plums, apricots, walnuts, olives and cherries? In the East we used to plant them in the fall, so as to have them freeze; as it does not freeze enough here, what do I have to do?

Do just the same. In California, heat and moisture cause the parting of the seed-cover, more slowly perhaps, but just as surely as the frost at the East. Early planting of all fruit pits and nuts is desirable for two reasons. First, it prevents too great drying and hardening and other changes in the seed, because the soil moisture prevents it; second, it gives plenty of time for the opening and germination first mentioned. But early planting must be in ground which is loamy and light rather than heavy, because if the soil is so heavy as to become water-logged the kernel is more apt to decay than to grow. Where there is danger of this, the seed can be kept in boxes of sand, continually moist, but not wet, by use of water, and planted out, as sprouting seeds, after the coldest rains are over, say in February. Cherry and plum seeds should be kept moist after taking from the fruit; very little is usually had from dry seeds. The other fruits will stand considerable drying. Very few olives are from the seed, because of reversion to wild types - also because it is so much easier to get just the variety you want by growing trees from cuttings.

Mailing Scions.

Which is the best way to send scions by mail?

Wax the ends of mature cuttings, remove the leaves and enclose in a tight tin canister with no wet packing material.

Nursery Stock in Young Orchard.

How will it do to raise, for two or three years, a lot of orange seedlings between the rows of young three-year-old orange trees? I see that a nurseryman near me has done this, and his trees are more flourishing than mine.

It can be done all right, as your own observation affirms. The superior appearance of the trees may be due to the additional water, and fertilizer probably, used to push the seedlings; possibly also to extra cultivation given them. It all depends upon what policy is observed in growing the seedlings; if something more than usual is done for their sakes, the trees may get their share and manifest it. If not, the trees will be robbed by the seedlings, and there is likely to be loss by both. There is no advantage in the mere fact that both are grown; there may be in the way they are grown. Whether there is money value in the operation or not depends upon how many undertake it.

Square or Triangular Planting.

What is your opinion on triangular planting as compared with square planting?

Planting in squares is the prevailing method. The triangular plan is not a good one when one contemplates removing trees planted as fillers. The orchard should either be planned in the square or quincunx form. In the latter case individual trees can be easily removed; in the other case rows can be removed - leaving the rows which you wish to keep equidistant from each other.

Killing Stumps by Medication.

Will boring into green stumps and inserting a handful of saltpeter kill the roots and cause the stump to readily burn up a few months later?

We have tried all kinds of prescriptions and have never killed a stump which had a mind to live. Many trees can be killed by cutting to stumps when in full growth, whether they are bored or not. Others will sprout in spite of all medicinal insertions we know of when these are placed in the inner wood of the stump. We believe a stump can be killed by sufficient contact with the inner bark layer of arsenic, bluestone, gasoline, and many other things, but it is not easy to arrange for such sufficient contact, and it would probably cost more than it would to blow or pull out the stump. One reader, however, assures us that he has killed large eucalyptus stumps by boring three holes in the stump with an inch auger, near the outer rim of the stump, placing therein a tablespoonful of potassium cyanide and saltpeter mixture (half and half), and plugging tightly. Another says: Give the stumps a liberal application of salt, say a half-inch all over the top, and let the fog and rain dissolve and soak down, and you will not have much trouble with suckers.

Planting Fruit Trees on Clearings.

We wish to plant orchard trees on land cleared this winter: manzanita and chaparral, but also some oaks and large pines and groves of small pines. We have been told that trees planted under such conditions, the ground containing the many small roots that we cannot get out, would not do well. Are the bad effects of the small roots liable to be serious; also, would lime or any other common fertilizer counteract the bad effects?

Proceed with the planting, as you are ready for it, and take the chances of root injury. It may be slight; possibly even absent. Carefully throw out all root pieces, as you dig the hole, and exclude them from the earth which you use in filling around the roots, and in the places where large trees stood, fill the holes with soil from a distance. Much depends upon how clean the clearing was. No considerable antiseptic effect could be expected from lime and the soil ought to be strong enough to grow good young trees without enrichment. The pear, fig and California black walnut are some of the most resistant among fruit-bearing trees, and these may usually be planted with safety. The cherry is the most resistant of the stone fruits. The "toadstool" disease occasionally affects young apple trees recently set out, but it is not usually serious on established trees.

Dipping Roots of Fruit Trees.

In planting an almond orchard would it be of any benefit to dip the young trees in a solution of bluestone and lime dissolved?

We doubt if it would serve any good purpose. If done at all the dip should be carefully prepared in accordance with the formula for bordeaux mixture, for excess of bluestone will kill roots. Healthy trees do not need such treatment, and we doubt if unhealthy ones can be rendered safe or desirable by it.

Preparing for Fruit Planting.

What effect will a crop of wheat have on new cleared land, to be planted in fruit trees later on?

One crop of wheat or barley will make no particular difference with the cleared land which you expect to plant to fruit later. It would be better to grow a cultivated crop like corn, potatoes, beets, squashes, etc., because this crop would require summer cultivation which would kill out many weeds or sprouts and leave your land in better shape for planting.

Depth in Planting Fruit Trees.

I have been advised to plant the bud scar above ground in a wet country.
Is that right?

On ordinary good loam, plant the tree so that it will stand about the same as it did in the nursery: a little lower, perhaps, but not much. The bud scar should be a little above the surface. It is somewhat less likely to give trouble by decay in the upset tissue. If the soil is heavy and wet, plant higher, perhaps, than the nursery soil-mark, but not much. In light, sandy soil, plant lower - even from four to six inches lower - than in the nursery sometimes. In this case the budscar is below the surface, but that does not matter in a light, dry soil which does not retain moisture near the surface.

Fruit Trees in a Wet Place.

One part of my orchard is low and wet, much scale and old trees loose. Will much spraying be a cure and can I use posts to hold the old trees firm, or would you take out and put in Bartlett pears!

Spraying would kill the scale but no spraying will make a tree satisfactory in inhospitable soil. As pears will endure wet places better than apples, it would seem to be wise to make the substitution, providing the situation is not too bad for any fruit tree. In that case you can use it for a summer vegetable patch.

Cutting Back at Planting.

I have planted a lot of one-year-old cherry trees and would like to know if I should cut them down the same as the apple tree? I have also planted a lot of walnut trees. Shall I cut them off?

Yes for the cherries and no for the walnuts - although we have to admit that some planters hold for cutting back the walnuts also. If you do cut back the walnuts, let them have about twice the height of stem you give the cherries and cover the exposed pith with wax or paint.

Branching Young Fruit Trees.

It is the practice in this locality to wrap all young trees to a point 24 inches above the bud, for the purpose of protection against rabbits, to protect the bark from the sun and to prevent growth of sprouts. These wrappings are kept on indefinitely, the rule being that no sprouting is to be permitted below the 24-inch murk. Is there any virtue in this, and why is it done?

The wrapping is desirable both to protect them from rabbits and from sunburn, and either this or whitewash or some other form of protection should certainly be employed against the latter trouble. It is not desirable to have all the branches emerge at the same point, either 24 from the ground or at some lower level, as is preferable in interior situations, but branches should be distributed up and down and around the trunk so as to give a strong, well-balanced, low-headed tree. So far as wrapping interferes with the growth of shoots in this manner it is undesirable.

Coal Tar and Asphaltum on Trees.

What is the effect of coal tar or asphaltum applied to the bark of trees?

The application of coal tar to prevent the root borers of the prune which operate near the surface of the ground was found to be not injurious to the trees, although there was great apprehension that there would be. The application of asphaltum, what is known as "grade D," has been also used to some extent in the Santa Clara valley without injury. Of course, in the use of any black material, you increase the danger of sunburn, if applied to bark which is reached by the sun's rays.

Whitewashing Fruit Trees.

When is the proper time to whitewash walnut trees to prevent sun scald?
How high up is it advisable to apply the wash?

Whitewash after heavy rains are over and before the sun gets very hot; near the coast see that it is on early in April; in the interior it should be in place in March. Do not wait until all the rains are over, because there is a great chance of bark-burning between rains in the spring. Whitewash the trunk and the larger limbs - wherever the sun can reach the bark; being careful to keep the surface white where the 2 o'clock sun hits it. Be particular to whitewash, or otherwise protect by "protectors" or burlap wrappings, all young trees; the young tree is more apt to be hurt than an old one, but bark seems never to get too old to burn if the sun is hot enough.

Shaping a Young Tree.

In shortening back long, slim limbs the side shoots come out, and one soon has a lot of ugly, crooked limbs to look at. There are a number of orchards here being spoiled in that way. How is this avoided?

You cannot secure a low-heading, well-shaped tree without cutting back the branches. Afterward you can improve the form by selecting shoots which are going in directions which you prefer, or you can cut back the shoots afterward to a bud which will start in the direction which you desire. In this way the progressive shaping of the tree must be pursued. If you only have a few trees and can afford the time, you can, of course, bend and tie the branches as they grow, so that they will take directions which seem to you better, but this is not practicable in orcharding on a commercial scale. There is no disadvantage in crooked branches in a fruit tree, but they should crook in desirable directions, and that is where the art in pruning comes in.

Pruning Times.

What is the best time to prune the French prune and most other trees? In Santa Clara volley they prune as soon as leaves are off; in the mountains they prune later, say in February and March, and finish after bloom is started and of course when sap is up. Which is right?

You can prune French prunes and other deciduous trees at any time during the winter that is most convenient to you. It does not make any particular difference to the tree, nor does it injure the tree at all if you should continue pruning after the bloom has started. In fact, it is better to make large cuts late in the winter, because they heal over more readily at the beginning of the growing period than at the beginning of the resting season. It is believed that early pruning may cause the tree or vine to start growth somewhat sooner and this may be undesirable in very frosty places.

Grafting Wax.

How shall I make grafting wax for grafting fruit trees?

There are many "favorite prescriptions" for grafting wax. One which is now being largely used in fruit tree grafting is as follows: Resin, 5 lbs.; beeswax, 1 lb.; linseed oil, 1 pint; flour, 1 pint. The flour is added slowly and stirred in after the other ingredients have been boiled together and the liquid becomes somewhat cooler. Some substitute lampblack for flour. This wax is warmed and applied as a liquid.

Plowing in Young Orchard.

How near can I plow to two-year-old orange trees safely?

You can plow young orange orchards as close to the trees as you can approach without injuring the bark, regulating depth so as not to destroy main roots. Destruction of root fibers which have approached too near the surface is not material. It is very desirable that the soil around and near the tree be as carefully worked as possible without injury to the bark of the tree. How far that can be done by horse work and how much must be done by hand must be decided by the individual judgment of the grower.

Crops Between Fruit Trees.

What would be best to grow between fruit trees, while the trees are growing, and what to alternate each season, so as not to use up the soil without putting back into it?

Where one is bringing along a young orchard, without irrigation, it is doubtful whether it is not better policy to give the trees all the advantage of clean cultivation and ample moisture than to undertake intercropping. If you live on the place and wish to grow vegetables between the rows, the thorough cultivation to bring the vegetables along satisfactorily would help to preserve moisture enough both for the vegetables and for the trees, but this is very different from growing a field crop by ordinary methods of cultivation. Select a crop which will require summer cultivation, like corn, potatoes, squashes, and beans, and never a hay or grain crop which takes up moisture without working the soil for the greater moisture conversation which hoed crops require. In choice of hoed crops be governed by what you can use to advantage, either for house or the feeding of animals, or what you can grow that is salable with least loss of moisture in the soil. The choice is governed entirely by local conditions, except that leguminous plants - peas, beans, vetches, clovers, etc. - do take nitrogen from the atmosphere and can thus be grown with least injury and sometimes with a positive benefit to the fertility of the soil.

Regular Bearing of Fruit Trees.

How can trees be induced to bear regularly instead of bearing excessively on alternate years?

The most rational view is that in order to bear regularly the tree must be prevented from overbearing by thinning of the fruit; also that the moisture and plant-food supply must be regularly maintained, so that the tree may work along regularly and not stop bearing one year in order to accumulate vigor for a following year's crop. There is some reason to believe that some trees which seem to overbear every year can be prolonged in their profitable life and made to produce a moderate amount of fruit of large size and higher value by sharp thinning to prevent overbearing at any time. This is found clearly practicable in the cases of the apricot, peach, pear, apple, table grape, shipping plum, etc., because the added value of larger fruits is greater than the cost of removing the surplus.

Scions from Young Trees.

I have bought some one-year-old apple trees that are certified pedigree trees. Would it be practical to take the tops of these trees and graft on one-year seedlings and get the same results as from the trees I bought? Will they bear just as good, or is it necessary to take the scions from old bearing trees?

They will bear exactly the same fruit as the young trees will, but you cannot tell how good that will be until you get the fruit. The advantage of scions from bearing trees is that you know exactly what you will get, for, presumably, you have seen and approved it.

Late Pruning.

Will I do injury to my peach trees if I delay pruning until the last of February, or until the sap begins to run and the buds to swell?

It will not do any particular harm to let your peach pruning go until the buds swell or even after the leaves appear. Late pruning is not injurious, but rather more inconvenient.

Avoiding Crotches in Fruit Trees.

How can I avoid bad crotches in fruit trees?

Crotches, which means branches of equal or nearly equal size, emerging from a point at a very acute angle, should be prevented by cutting out one or both of them. The branching of a lateral at a larger angle does not form a crotch and it usually buttresses itself well on the larger branch. That is a desirable form of branching. Short distances between such branchings is desirable, because it makes a stronger and more permanently upright limb, capable of sustaining much weight of foliage and fruit. Build up the young tree by shortening in as it grows, so as to get such a strong framework.

Crotch-Splitting of Fruit Trees.

I have a young fig tree that is splitting at the crotches. I fear that when the foliage appears, with the force of the winds the limbs will split down entirely.

Perhaps you have been forcing the trees too much with water and thus secured too much foliage and weak wood. Whenever a tree is doing that, the limbs ought to be supported with bale rope tied to opposite limbs through the head, or otherwise held up, to prevent splitting. If splitting has actually occurred, the weaker limb should be cut away and the other staked if necessary until it gets strength and stiffens. If the limbs are rather large they can be drawn up and a 3/16 inch carriage bolt put through to hold both in place; but this is a poor way to make a strong tree. We should cut out all splits and do the best we could to make a tree out of what is left. Then do not make them grow so fast.

Strengthening Fruit Trees.

I have read that some trees are propped by natural braces; that is, by inter-twining two opposite branches while the tree is young, so that in time they grow together. What is your idea regarding the practicability of such an idea in a large commercial orchard?

Twining branches for the purpose indicated is frequently commended, but it seems best for the use of ingenious people with plenty of time and not many trees. To prune trees to carry their fruit so far as one can foresee, and to use props or other supports when a tree manifests need of a particular help which was not foreseen is the most rational way to handle the proposition on a large commercial scale.

Time for Pruning.

What is the proper time for pruning pear and apricot trees?

Ordinary deciduous fruit trees can be successfully pruned from the time the leaves begin to turn yellow and fall, until the new foliage is appearing in the late winter or spring.

Grape Planting.

What is the proper time for planting grape vines?

Grape vines are most successfully planted after the heavy rains and low temperatures are over and before the growth starts: This will usually be whenever the soil is in good condition, during the months of February and March.

Covering Tree Wounds.

What is the best stuff to use on wounds and large cuts on my fruit trees? I have used grafting wax, but it is expensive and not altogether satisfactory.

Amputation wounds on trees can be more successfully treated with lead and oil paint than with grafting wax. Mixed paint containing benzine would not be so good as pure lead and oil mixed for the purpose and then carefully applied as to amount so as not to run. "Asphaltum Grade D" may also be used in the same way.

Covering Sunburned Bark.

Would asphaltum do to use an sunburned bark?

Owing to the attraction of the heat by the black color, asphaltum would increase the injury by absorption of more heat. Some white coating is altogether best for sunburn injuries, because it will reflect and not absorb heat, and a durable whitewash applied as may be needed to keep the white covering intact is undoubtedly the best treatment. Where the bark has been actually removed, white paint would be superior to whitewash to keep the wood from checking while the wound was being covered laterally by the growth of new bark.

Too Much Pruning.

Same peach trees entering the third year were pruned early in the winter very severely. The pruner merely left the trunk and the three or four main laterals, the latter about one foot in length. A large proportion of these trees have not sprouted as yet, though alder and better pruned trees are all sprouted in the same vicinity. The bark is green and has considerable sap. Will the trees commence to grow?

The trees will sprout later, after they have developed latent buds into active form. The pruning probably removed all the buds of recent growth. After starting they will make irregular growth, starting too many shoots in the wrong places, etc., and considerable effort will be necessary to get well-shaped trees by selection of shoots in the right places and thinning out those which are not desirable.

For Broken Roots.

When the root of an orange or other fruit tree is exposed or brakes by the cultivator, what is the best way to treat that root?

Where a root is actually broken it is best to cut it off cleanly above the break. This will induce quick healing over and the sending out of other roots. Where there is only a bruise on one side, all the frayed edges of the wound should be cleanly cut back to sound bark, which will have a tendency to promote healing and prevent decay.

Pruning in Frosty Places.

This appears to be a frosty section. Pruners are at work continuously from the time the apricots are harvested until spring arrives. From what is said in "California Fruits?" I judge late winter pruning would be best far apricots and peaches. Am I correct?

In frosty places it is often desirable to prune rather late, because the late-pruned tree usually starts later than the early pruned, and thus may not bloom until after frost is over.

Low Growth on Fruit Trees.

Should the little twigs an the lower parts of young fruit trees be removed or shortened?

An important function which these small shoots and the foliage which they will carry perform is in the thickening of the larger branches to which they are attached and overcoming the tendency of the tree to become too tall and spindling. This can be done at any time, even to the pinching of young, soft shoots as they appear. It must be said, however, that in ordinary commercial fruit growing little attention is paid to these fine points, which are the great enjoyment of the European fruit-gardeners and are of questionable value in our standard orcharding. It is, however, a great mistake to clear away all low twigs, for such twigs bring the first fruit on young trees.

Are Tap-Roots Essential?

Is it better to plant a nut or seed or to plant a grafted root; also is it better to allow the tap-root to remain or not in event of planting a grafted root?

It does not matter at all whether the tree has its original tap-root or not. All tap-roots are more or less destroyed in transplanting and the fact that not one per cent of the walnut trees now bearing crops in California consist of trees grown from the nut itself planted in place, is sufficient demonstration to us that it is perfectly practicable to proceed with transplanting the trees. It is more important that the tree should have the right sort of soil and the right degree of moisture to grow in than that it should retain the root from which the seedling started. The removal of the tap-root does not prevent the tree from sending out one or several deep running roots which will penetrate as deeply as the soil and moisture conditions favor. This is true not only of the walnut but of other fruit trees.

Transplanting Old Trees.

Can I transplant fruit trees 2 to 3 inches through the butt, about one foot from the ground? Varieties are oranges, lemons, pears, apples and English walnuts nearly 4 inches through the butt. I wish to move them nearly a mile. What is the best way and what the best month to do the work, or are trees too large to do well if moved?

The orange and lemon will do better in transplanting than the others. Take up the trees when the soil becomes warmed by the sun after the coldest weather is over. This may be in February. Cut back the branches severely and take up the trees with a good ball of earth, using suitable lifting tackle to handle it without breaking. Settle the earth around the ball in the new place with water, and keep the soil amply moist but not wet. Whitewash all bark exposed to the sun by cutting back. You can handle the walnut the same way, but it would, however, probably get such a setback that it might be better to buy a new tree two or three years old and plant that. The apples and pears we would not try to transplant, but would rather have good new yearlings than try to coax them along. Transplanting deciduous trees should be done earlier in the winter than evergreens.

Dwarfing a Fruit Tree.

I am told that by pruning the roots of a young tree after the root system is well started (say three years old) that as a result this will produce a tree that is semi-dwarfed or practically a dwarfed fruit tree.

Yes; cutting back the roots in the winter and cutting back the new growth in the summer will have a dwarfing effect. The best way to get a dwarfed garden tree is to use a dwarfing root. You can get trees on such roots at the nurseries.

Seedling Fruits.

I have been growing seedlings from the pits of some extra fine peaches and plums with a view to planting them. A man near San Jose advised me that I would get good results, but since then I have met others who say that the fruit trees that spring from planted seeds yield only poor fruit.

It is the tendency of nearly all improved fruit to revert to wild types, more or less, when grown from the seed. The chances are, then, that nine-tenths or more of the seedlings which you grew for fruiting might be worthless. A few might be as good as the fruit from which you took the pits; possibly one might he better. For these reasons the growing of fruit trees from pits and seeds is only used for the purpose of getting a root from which a chosen variety may be gotten by budding and grafting.

Grafting.

I did a little grafting last spring, and as it was my first attempt, about ten per cent of the scions failed to grow. Now shall I saw the stub off lower down and try again, or bud into one of the sprouts that have grown around the cut end? The trees are pear and cherry.

You did very well as a beginner not to lose more than one-tenth. Saw off below and graft again. You might have budded into one of those shoots last July, and if you fail again, bud into the new shoots next summer.

Filling Holes in Trees.

I have a number of trees that, on account of poor pruning and improper care, are decaying in the center. Many of them are hollow for a foot or more down the trunk.

Excavate all the decayed wood with a chisel or gouge or whatever cutting tool may work well and fill the cavity with Portland cement in such a way as to exclude moisture. This will prolong the life and productiveness of the trees for many years if other conditions are favorable.

Deferring Bloom of Fruit Trees.

Have any experiments ever been carried on definitely to decide what causes early blossoming of fruit trees? For instance, have adjacent trees of the same variety been treated definitely by putting a heavy mulch around one to hold the cold temperature late in the spring, leaving the other tree unmulched so the roots could warm up?

It has been definitely determined by the experiments of Professor Whidden of the Missouri Experiment Station that the swelling of the buds and starting of the foliage of fruit trees is due to the action of heat upon the aerial parts of the trees; that is, growth is not caused by increasing the temperature of the ground and cannot be retarded by cooling the ground. Experiments with the use of snow and ice under trees by which the ground has been kept at a low temperature have not prevented the activity of the tree. The only way known to retard activity is to spray the tree with whitewash so that the white color may reflect the heat and prevent the absorption of it by the bark, which is usually of a dark color and therefore suited to heat absorption. Retarding of growth is possible in this way for a period of six to ten days, which, of course, in some cases might be of value, but the lengthened dormancy is probably too small to constitute it of general value. In whitewashing, to determine what advantage there is in it in retarding growth, the tree should be thoroughly sprayed with whitewash so as to cover all the wood some time before the buds swell. In fact, it is to prevent the early swelling of the buds that the whitewashing is resorted to. It is better to make the application, therefore, a little too early than too late. A specific date cannot be given for it that would be right in all localities.

Repairing Rabbit Injuries.

Your book says in Pruning young trees for the first time, about four main branches should be left and these cut back to 10 or 12 inches. Now, where the rabbits have pruned back to 4 or 5 inches the very ones I wanted, what should be done? Some say, cut these back to the stem, allowing new shoots to start from the base of branches so removed.

Cut back to a bud near the stem, or if you do not see any, cut back near to the stem, but not near enough to remove the bark at the base of the shoot, for there are the latent buds which should give you the growth. This should be watched, and the best shoot selected from each point to make a strong branch, pinching back or removing the others.

For a Bark Wound.

What is best to do with an apricot or prune tree when it has been hit with an implement and the bark knocked off?

Cut around the bark wound with a sharp knife so as to remove all frayed edges. Cover the exposed wood with oil and lead paint to prevent cracking, and the wound will soon be covered with new bark from the sides.

Bridging Gopher Girdles.

How shall I make the bridge-graft or root-graft over the trunks of trees girdled by gophers? Has this method proved successful in saving trees three or four inches in diameter, and how is it done?

The bridging over of injury by mice by grafting has been known to be successful for decades in countries where this trouble is encountered. Undoubtedly the same plan would work in the case of all bark injuries which can be bridged. The plan is to take good well-matured shoots which are a little longer than the injury which has to be spanned, making a sloping cut on both ends, also a cut into the healthy bark above and below the injury, and slip the cut ends of the cutting into the cuts in the bark so that the ends go under the bark above and below, and the cut ends are closely connected with the growing layer of the stock. If the cutting is made a little longer than the distance to be spanned, the tendency of the cutting by straightening is to hold itself in place. When in place, the connections should he covered with wax to prevent drying out.

Soil-Binding Plant for Winter.

What would be the best to plant in an orchard on ground of a light sandy sediment which, after plowing, will move with the strong winds? I would like to plant something that will benefit the ground. The winds are the strongest from December to April. This is in the irrigated district and I need something that will make a sod during that period.

We would, in all the valleys, advise a fall irrigation (if the rains are late) and the sowing of burr clover, which when started in September will have the ground well covered by December, if you keep the moisture right to push it. Disking or plowing this under in March (or April, according to locality) will hold the sand and afterward enrich it. You can do this every year, but probably you will not need to seed it more than once.

Bananas in California.

Is there any reason why bananas would not grow and bear in the vicinity of Merced if they had plenty of water? Or would the cool nights at certain seasons keep them from bearing? Would they do better in the Imperial valley?

Bananas would suffer too severely from frost to be profitable at any point in the interior valleys of California. A plant would be killed to the ground at least every year unless under glass or other protection. There are a few places practically frostless where bananas can be grown in this State, but there is no promise in commercial production because they can be so cheaply imported from the tropics.

Carobs in California.

Will the carob tree (St. John's Bread) do well in the Sacramento valley, and is it a desirable tree for lining a driveway?

Carobs have been grown in California for thirty years or more and they will make a handsome driveway and give a lot of pods for the kids and the pigs - for they are "the husks which the swine did eat," and both like them. They ought to be much more widely planted in California because they grow well and are good to look upon.

Spineless Cactus Fruit.

I have about two acres of high land in Fresno county that can't be irrigated. It is red adobe soil and there is hardpan in it. Which kind of fruit trees will grow and pay best? How near may the hardpan be to the surface before I have to blast it?

It is a hard fruit proposition. Try spineless cactus, the fruits of which are delicious. Blasting would help if there is a moist substratum below the hardpan and might enable you to grow many fruits. If your land is hard and dry all the way down, blasting would not help you unless you can get irrigation. Presumably your rainfall is too small for fruit unless you strike underflow below the hardpan.

Cleaning Fruit Trays.

What do you advise for killing and removing the whitish mold that forms on trays used for drying prunes? Would sunning the trays be effective, or washing in hot water, or is there some suitable fungicide?

Good hot sun and dry wind will kill the mold. The spores of such a common mold are waiting everywhere, so that your fruit would mold anyway if conditions were right. Still, scalding the trays for cleanliness and a short trip through the sulphur box for fungus-killing is commended.

Killing Moss on Old Trees.

I have some Bartlett pear trees that are covered with moss and mold, and the bark is rough and checked. I have used potash (98%), 1 pound to 6 gallons spray. It kills the long moss, but the green mold it does not seem to affect. The trees have been sprayed about one week. Some trees have been sprayed with a 1 pound to 10 gallons solution by mistake. Shall I spray these again with full strength, and when?

You have done enough for the moss at present. Even the weaker solution ought to be strong enough to clean the bark. Wait and see how the bark looks when the potash gets through biting; it will keep at it for some time, taking a fresh hold probably with each new moisture supply from shower or damp air. The spray should have been shot onto the bark with considerable force - not simply sprinkled on.

Shy-Bearing Apples.

I have some apple trees 10 and 12 years old that do not bear satisfactorily, but persist in making 5 to 6 feet of new wood each year. If not cut back this winter, will they be more likely to make fruit buds?

Yes, probably. Certainly you should try it. You should also cultivate less and slow down the growth. If they then take to bearing, you can resume moderate pruning and better cultivation. This is on the assumption that your trees are in too rich or too moist a place. But you should satisfy yourself by inquiry and observation as to whether the same varieties do bear well in your vicinity when conditions are such that slower growth is made. If the variety is naturally shy in bearing, or if it requires cross-pollination, the proposed repressive treatment might not avail anything. In that case you can graft over the tree to some variety which does bear well or graft part of the trees to another variety for cross-pollination.

No Apples on Quince.

How large a tree will the Yellow Bellefleur apple make if grafted or budded on quince root at the age of 15 years? I have been trying to get some information about dwarf fruit trees, but it is difficult to get.

No wonder the information is hard to get. The Yellow Bellefleur will not grow upon the quince at all, or at least not for long. In growing dwarf apples the Paradise stock is used, while the quince is used for dwarfing the pear, and many varieties of pears will accept the quince root which the apple rejects.

Stock for Apples.

Do you recommend French seedling stock as greatly to be preferred to that grown in this country?

French seedling stock is generally used because it is graded and furnished in uniform sizes; also, because it can usually be purchased for less than seedlings can be grown under our labor conditions. Locally grown apple seedlings are apt to be irregular in size and, as already stated, cost more than the properly graded imported stock.

Apples and Alfalfa.

I have recently come across a proposition to sow apple orchards in the interior of southern California with alfalfa. The apples are said to be superior and the crop heavier, to say nothing of a half or two-thirds of an alfalfa crop in addition to the crop of apples. What do you know about it? Is alfalfa being used by others in this way?

It is perfectly rational to grow alfalfa in fruit orchards if the water supply is ample for both the trees and the intercrop and the owner will not yield to the temptation to waterlog his trees for the sake of getting more alfalfa. It is even more desirable in the interior than near the coast, probably. In Arizona some growers have for a number of years practiced growing alfalfa in orchards, cutting the alfalfa without removing it, counting that clippings are worth more to them through their decay and the increase of the humus content of the soil. Even where this is not done, the alfalfa will add to the humus of the soil by its own wastes both from root and stem. The presence of an alfalfa cover reduces the danger of leaf and bark burning either by reflected or radiated heat from a smooth ground surface, and some trees are very much benefited by this protection in regions of high temperature. This might be expected to be the case with the apple, which is somewhat subject to leaf burning in our interior valleys.

Top Grafting.

In grafting over apple and pear trees to some other variety, is it advisable to cut off and graft the entire tree the first year where the trees are from 7 to 15 years old, or would it be better to cut off only a part of the top the first year and the rest the following year?

In the coast region it is a good practice to graft over the whole tree at one time, cutting, however, above the forks and not into the main stem below the forking. This gives many scions which seem able to take care of the sap successfully. In the interior valleys, it is rather better practice to leave a branch or two, cutting them out at the following winter's pruning, for probably the first year's grafts will give you branches enough. This has the effect of preventing the drowning out of the scions from too strong sap-flow. Cutting back and regrafting of old trees should be done rather early, before the most active sap-flow begins. The later in the season the grafting is done, and the warmer the locality, the more desirable it seems to be to leave a branch or two when grafting.

Apple Budding.

What is the best time to bud apples?

Apples are budded in July and August and remain dormant until the following spring.

Mildew on Apple Seedlings.

Why do young apple plants in the seed bed became mildewed? They are in a lath house.

Because too much moisture was associated with too much shade. More sunshine would have prevented mildew, and if they had enjoyed it the seedlings could have made better use of the water probably.

Pruning Apples.

Young apple trees set two years ago were cut back to 14 to 18 inches and cared for as to low branching, proper spacing, etc., but the desired branches were allowed to make full growth to the present time. They have mode great growth and if allowed to continue will make too tall trees.

We understand that your trees have made two summers' growth since pruning. We should cut back to a good lateral wherever you can find one running at the right direction at about three to four feet from the last cut, and shorten the lateral more or less according to the best judgment we could form on sight of the tree. In this way you can take out the branches which are running too high and make the framework for a lower growth. Do not remove the small twigs and spurs unless you have too many such shoots.

Cutting Back Apples and Pears.

"California Fruits" says the "apple does not relish cutting back, nor is it desirable to shorten in the branches." But when a three-year-old tree gets above 12 feet high, as many of mine are doing, what are you going to do? I cut these back same last year, but up they go again with more branches than ever. The pears are getting too tall, also. Should not both apple and pear trees be kept down to about ten feet?

The quotation you make refers to old bearing trees, and indicates that their pruning is not like that of the peach, which is continually shortened in to keep plenty of new wood low down. Of course, in securing low and satisfactory branching on young apples, pears, etc., there must be cutting back, and this must be continued while you are forming the tree. If you mean that these trees are to be permanently kept at ten feet high, you should have planted trees worked on dwarfing stocks. Such a height does not allow a standard tree freedom enough for thrift; as they become older they will require from twice to thrice the altitude you assign to them, probably. Pears can be more successfully kept down than apples, but not to ten feet except as dwarfs.

Pruning Old Apple Trees.

How would you prune apple trees eight or nine years old that have not been cut back? There are a great many that have run up 20 feet high with twelve or fifteen main limbs and very few being more than two or three inches in diameter.

Remove cross branches which are interfering with others and thin out branches which seem to be crowding each other at their attachments to the trunk, by removing some of them at the starting point. Having removed these carefully so as not to knock off spurs from other branches, study the tree as it is thus somewhat opened up and see where remaining branches can be shortened to overcome the tendency to run too high. Do not shear off branches leaving a lot of stubs in the upper part of the tree, but always cut back a main branch to a lateral and shorten the lateral higher up if desirable. This will keep away from having a lot of brush in the top of the tree. Study each tree by itself for symmetry and balance of branches and proceed by judgment rather than by rules anyone can give you.

Top-Grafting Apples.

Can I graft over a few Ben Davis apple trees 25 years old or there-abouts, but thrifty and vigorous?

It is certainly possible, by the old top-grafting method which has been used everywhere with apples for centuries. Graft during the winter. Work on the limbs above the head so as to preserve the advantage of the old forking, using a cleft graft and waxing well. It is usually best to graft over a part of the limbs and the balance a year later.

Will the Apples Be the Same Kind?

I have a mixed orchard, mostly Gravensteins, and I want to graft all the other trees into a Gravenstein top if I can do so and at the same time get the early Gravenstein bloom and the fruit would be as satisfactory as though on other roots.

The new tree grown from the grafts will behave just like the tree from which the scions were taken if similarly thrifty.

Places for Apples.

What quality is it in the soil in the vicinity of Watsonville that makes that country peculiarly adapted to the culture of apples? Are there not other portions of the State where apples could be produced on a commercial basis?

It is not alone quality in the soil, but character of the climate that underlie success in the Watsonville district. Apples can be and are grown on a commercial scale through the coast district of Sonoma, Mendocino, and Humboldt counties; also in suitable situations in the coast counties south of Santa Cruz county. Along the coast, as far as deep retentive soil and the cool air of the ocean extend, one may expect to get apples similar to those produced in the Watsonville district. In the interior valleys, on suitable soils with adequate moisture, early apples are profitably grown, while in the higher foothill and mountain valleys in all parts of the State, where moisture is sufficient, late keeping apples of high quality are produced.

Summer-pruning Apples.

Will summer pruning cause apple trees to bear fruit instead of growing so much new wood?

Over-growth can be repressed by summer pruning, and if done just at the right time bearing is increased and late new growth is avoided, but it is not easy to determine exactly the right time, and it has to be fixed according to local conditions of length of growing season and growth condition of the tree itself also. It is better for some varieties than others, and, in fact, has to be done wisely. A summer slashing of apple trees, simply because some one says so, is not only expensive, but may do more harm than good. Therefore, those inclined to it, should try a few trees at first and note results.

Grafting Apple Seedlings in Place.

I want to plant apple trees for home use. I have an idea to plant apple seeds instead of trees: planting three or four seeds for each hill, right in the place where I would grow the trees, and select the best one to graft on. I will take seed of Bellefleurs, which are vigorous growers. What do you think? Will the seed germinate readily and when is the right time to plant?

Select plump, well ripened seed, keep them in damp sand until the ground begins to get warm in January or February, according to location. But such an undertaking will cost you vastly more in time, in labor, and waste of land than it would to buy well-grown nursery trees budded with the variety which you desire. Such trees would give you practically a uniform lot of trees in your orchard while planting seedlings and grafting afterward would give you very irregular and for the most part unsatisfactory results - providing you get any seeds to grow at all in the open ground, which is doubtful.

Resistant Apple Roots.

A few apple trees which are almost dead from ravages of the woolly aphis. I am going to dig them out and plant in their places other apple trees on woolly aphis-proof root. Will it be necessary to use measures to exterminate the woolly aphis in the old roots or their places in the ground before planting new trees in the places of the removed trees?

It is not necessary to undertake to kill aphis in the ground when you are planting apple trees on resistant roots. It will give your trees a better start to dig large holes, throw out the old soil, and fill in with some new soil from another part of the land to be planted, but it has been demonstrated that these roots are resistant, no matter if planted in the midst of infestation.

Apples and Cherries for a Hot Place.

What kind of apple do you think would do best in a dry, hot climate?
What do you think of the Early Richmond cherry in such a place?

Apples most likely to succeed in a dry situation are those which ripen their fruit very early. The Red Astrachan is on the whole the most satisfactory, but there are many places which are altogether too dry and hot for any kind of apple. Whether cherries would succeed or not you can only tell by trying. Possibly the trees would not live through the summer if your soil becomes very dry. The most hardy cherries are the sour or pie cherries and the Early Richmond is one of this group.

Die-back of Apple Trees.

What causes the death of the top shoots in apple trees?

New wood is sometimes diseased by mildew, but die-back is usually due to two different causes: One, the accumulation of water in the soil during the excessive rains of mid-winter; second, the occurrence of low temperatures, including frosts, after the sap has risen. Which of these causes operate in a certain case depends, of course, upon whether the soil was heavy and inclined to retain standing water too long, or whether there were such frosts at about the time when the leaves should start. Sometimes, of course, both of these conditions worked in the same place; sometimes one and sometimes the other, but certainly both of them are capable of causing the trouble. There seems to be no specific disease; it is rather a matter of unfavorable conditions for growth.

Storage of Apples.

We desire to store two or three thousand boxes of apples for three or four months and propose to do it in this way: Make an excavation in dry earth, putting at the bottom of the excavation straw. Upon this straw place the apples, then dry straw over the apples, and upon the top of this two or three feet of dry earth. Will it be a good plan to pour on water from time to time over the top of this to keep the apples and all wet, or should the apples be kept dry?

Putting down loose apples in a straw-lined pit would be very expensive. It would invite decay by bruising the fruit, and the result would probably be a worthless mixture of rotten fruit and straw. The fruit should be stored in boxes or shallow trays to reduce pressure and promote ventilation, and not in bins or large piles. Apples will keep for a long time in good condition if the boxes are put in piles in the shade, covered with straw, which should be slightly moistened from time to time; but in that case there would not be such an accumulation of moisture and there would be ventilation at all times. Apples should be kept dry, but they will shrivel and become unmarketable unless the air in which they are stored is kept reasonably moist. This is generally accomplished by making apple houses with double walls and roof to exclude heat and with an earth or concrete floor which can be sprinkled from time to time with a hose.

Apple Root-grafts.

I have an old apple orchard and would like to have two or three of the best varieties positively identified, so that I can order these kinds from the nursery for next year's planting.

Old California apple orchards have many varieties no longer propagated largely. If you greatly desire to have a few trees of exactly the varieties which you are now growing, you run some risk of mistake in ordering by name, but if you make some root-grafts by taking a piece of the smaller roots of the tree, which you can dig out, say about the size of a pencil, and graft scions upon them, you can secure root-grafts for planting in nursery this year and in that way be sure to have trees of exactly the same kind. Root-grafts can be made in the winter, placed in sand which is kept moist and not wet, planted out as soon as the ground warms up, and you will get immediate and very satisfactory growth in that way.

Pruning Old Apple Trees.

I have an old orchard containing some apple trees about 40 years old - trees well shaped but with plenty of main branches and limbs all very long. The trees bear profusely in alternate years but the fruit is small. In pruning would you advise cutting out some main limbs where there are over three or four and thus making a big wood reduction (where sunburn protection can still be guarded) or would you only shorten in the branches and thin the fruit severely?

Do not remove main branches unless they are clearly too numerous or have been allowed to grow to interference with each other or have become weakened or feeble in some way. In such cases the space is worth more than the branch. If the tree has a fair framework do not disturb it in order to get down to an arbitrary limit of three or four main branches; sometimes the tree can carry more. If the tree is too thick, thin it out by removing side branches of more or less size - saving the best, judging by both vigor and position. Work through the whole top in this way until you reach the best judgment you can form of enough space and light for good interior foliage and fruit. Apple branches should seldom be shortened, and when this seems desirable, cut to a side branch and not to a stub which will make a lot of weak shoots or brush in the top of the tree.

Pruning Apple Trees.

There is a great difference of opinion here regarding the pruning of three-year or older apple trees. Many people cut back three, four and five-year-old trees half the season's growth; others only cut back six inches.

Apple trees are cut back during their early life to cause branching and to secure short distances between the larger laterals on the main branches. This secures a lower, stronger tree. Cutting back twice or three times should secure a good framework of this kind, and then the apple should not be regularly and systematically cut back as the peach and apricot are. It is not possible to prescribe definite inches, because cutting back is a matter of judgment and depends upon how thick the growth is, what its position and relation to other shoots, etc. The chief point in cutting back is to know where you wish the next laterals to come on the shortened shoot, and if you do not wish more laterals at once; do not cut back at all. Treatment, of laterals which come of themselves is another matter. Do not clip the ends of shoots unless laterals are desired. If you keep clipping the ends of apple twigs, you will get no fruit from some varieties.

Grafting Almond on Peach.

I had good success with the peach trees which I grafted to almond last spring, getting about 95 per cent of a stand, and many of the grafts now are one and one-half inches diameter. In each of the trees I left about a quarter of the branches, to keep up the growing process of the tree. The universal practice around here in grafting is to cut the whole top off the tree at the time of grafting, but the increased growth and vigor of the grafts I have has proved to me and other growers around, that much better results are obtained by leaving part of the top on the tree at the time of grafting.

You did exceedingly well with your grafting. It seems a more rational way to proceed than by a total amputation, and yet ample success is often attained by grafting for a whole new top at once.

Pruning Almonds.

Should the main branches be shortened in a three-year-old almond tree? Of course, I intend to thin out the branches. Some growers here advise me to shorten the main branches; others say do not shorten them, as it tends to give the trees a brushy top.

Although some growers are contending for regular shortening - in of the almond as is practiced on the peach, it is not usual to cut back almond trees after they have reached three years of age and have assumed good form. Of course, if cutting back is done, the shoots coming from near the amputation must be thinned out to prevent the brushiness your adviser properly objected to.

Budding and Grafting Almonds.

Is it better to bud or graft bitter almond seedlings of one year's growth, and, as they must be transplanted, would it be proper to do the work this season or defer it for another year's growth?

Your almond seedlings should have been budded in July or August after starting from the nut, which would have fitted them for planting in orchard the following winter as dormant buds, as they cannot stay where they are another season. Now you can transplant to nursery rows in another place: cut back and graft as the buds are swelling, allowing a good single shoot to grow from below on those which do not start the grafts into which you can bud in June, and cut back the stock to force growth as soon as the buds have taken. In this way you will get the whole stock into trees for planting out next winter. Some will be large and some small, but all will come through if planted in good soil and cared for properly. Of course, you can plant out the seedlings and graft and bud in the orchard, but it will be a lot of trouble and you will get very irregular results.

Cutting Back Almonds.

I have some nice thrifty two-year-old almond trees which I did not "top" this spring. The limbs are from about four to seven or eight feet long. Would it not be best to "top" them yet?

Cut them back to a shoot of this year's growth, removing about a third of last year's growth, perhaps. This will give you lower and better branching.

Almond Planting.

I am contemplating the planting of about five to eight acres of almonds: what variety is best to plant?

Before planting so many almonds, you should determine how satisfactory the almond is in bearing in your location. Unless you can find satisfactory demonstration of this fact, it is hazardous to plant such an acreage. On the other hand, if you find that almonds are bearing satisfactorily, the kinds which are perhaps most satisfactory to plant are Nonpareil, Texas Prolific, Ne Plus Ultra and Drake's Seedling. The Texas Prolific and Drake's Seedling are abundant bearers and profitable because of the size of the crop, although the price is lower than the soft-shelled varieties, Nonpareil and Ne Plus Ultra. These two varieties are such energetic pollinizers that they not only bear well themselves, but force the bearing of the larger varieties mentioned. Every third row in your plantation should be either Texas Prolific or Drakes' Seedlings, which would give you two-thirds of the larger varieties and one-third of the smaller. There are, of course, other soft-shelled almonds which are worth planting and are being considerably planted in localities where they do well. This you can ascertain by inquiry among local growers and nurserymen. The planting of a good proportion of active pollinizers is the most important point.

Almond Pollination.

My almond trees look healthy but the fruit seems to be diseased. Is it necessary to have male and female trees, and how can one distinguish them?

The almond is monoecious and has perfect blossoms, therefore, there is no such thing as male and female trees in the case of the almond, but most of the best soft-shelled almonds are self-sterile and need cross-pollination from another variety. This is discussed elsewhere in answer to another question.

Roots for the Almond.

Which is the best root to have the almond grafted on, peach or bitter almond? The soil is sandy.

The bitter almond and the hard-shelled sweet almond are both used and we are not aware that any particular advantage has been demonstrated for either of them. The almond does well on peach roots also, but the almond is a better root where the soil conditions suit it.

Longevity of Almond and Peach.

What is difference in life of peach and almond in California?

The almond is the longer-lived, but we have seen both assuming the aspect of forest trees in abandoned pioneer places. Both are apt to live longer than their planters, if soil and moisture conditions favor.

Almond Seedlings.

I have been told that almond trees raised from seed, no matter what kind of seed planted, will produce bitter almonds. Is this a fact?

It is not a fact. The majority will probably be hard-shell, sweet and bitter, but others will be soft-shell, medium-shell, paper-shell, and everything else you ever heard of in the almond line. The almond has the sportiest kind of seedlings.

Do Not Plant Almonds in Place.

I have 30 acres which I intend to plant to almonds and peaches, and I thought of planting the sprouted nuts and pits where I wanted my trees, and budding the same there in orchard form. As one or two years' use of the land is not considered, what is your advice? My idea is to plant in orchard at start so as not to disturb roots, as when grown in nursery and transplanted in orchard. Would it not progress as rapidly? Would you advise budding peaches on almond roots; if not, why? My idea is that it would give a longer-lived tree.

We would do nothing of the kind. If we decided it better to grow trees than to buy them, we would grow and bud the seedlings in nursery and not in the field. Field budding is open to all kinds of injuries and growth from it, when saved from cultivation and all kinds of intruders, is irregular and uncertain. As for starting the roots from the nut in plate, it is largely a fanciful consideration. We count it no gain for the walnut which makes a tap root, and still less gainful for the almond and peach, which, usually make spreading roots. To cut off a tap root does not prevent the tree from rooting deeply if the soil is favorable. As to use of the land, you lose time by growing the seedlings in place. The peach does well on the almond root if soil conditions favor the almond. Perhaps it gives longer life to the peach, but the profitable life of the peach tree in a proper soil does not depend on the root; it depends upon the treatment of the top in pruning for renewal of branches.

Almond and Peach.

With water-table at 18 feet, which root is best for almond trees? The experience around here is that the peach root starts best. Which root is most durable? What is the life of the peach root and of the almond?

It is not merely a question of depth to water, but of character of the soil above the water. Neither of the roots will stand heavy soil which holds water too long, and both enjoy a free loam which drains readily down to the water-table or bottom water. If the soil is rather sandy, letting the water down very quickly, the almond is better in getting to it than the peach. If it is finer and still well drained the peach will do well, and the almond enjoys that also. The almond probably can be counted on to stand coarser soil and greater drouth than the peach and under such conditions will outlive the peach, probably, but both of them will live twenty to thirty years or more if pruned in the head to get enough new wood and the trunk is kept from sunburn. Aside from this choose the almond root for the almond.

Pollination of Almonds.

I have Drake's Seedling almonds. Some people have told me that I must plant some hardshell variety between them, otherwise they will not bear.

It is not necessary to plant hardshell almonds near Drake's Seedling trees in order to have them bear. Some varieties of almonds will set few nuts unless they are cross-pollinated, but these are the paper-shell varieties, as a rule - the Nonpareil, IXL and Ne Plus Ultra - and for these the Drake's Seedling or Texas Prolific is planted as a pollenizer. The highest-priced nut of all is the Nonpareil, and it is also a good bearer when in a good location and planted with Drake's or Texas Prolific.

Stick-tight Almonds.

I have leased seven acres of bearing almond trees which have the appearance of being reasonably well cared for. I notice a few trees that still have almonds on ("stick-tights"). What is the cause and remedy?

The occurrence of stick-tights is generally due to lack of moisture and thrifty growth, although some trees may be weak from some other cause and therefore deficient in sap-flow, which manifests itself in that way. Single nuts may also fall into that condition of malnutrition. We know no remedy except to keep the trees in good thrift by cultivation or by the use of irrigation if necessary.

Shy-bearing Apricots.

Why do my apricot trees not bring fruit? They seem healthy and are vigorous-looking trees. Five large trees have not borne 100 pounds of fruit in three years. The trees are not over six years old.

You may have a shy-bearing kind of apricot, of which there are many, or the trees may have grown too fast to hold the fruit, or the frost or north wind may have blasted the bloom. Stop winter pruning, and summer prune to prevent excessive growth; reduce irrigation; try to convince the apricot that it is not a "green bay tree" and see what will happen.

Pruning Apricots.

In pruning apricots, if there should be a hollow center of a big branch in center of a seven-year-old tree, should it be cut out with summer pruning? Should heavy growing apricots be summer pruned? Would it be all right to thin out a dense growth of wood in the prune trees in September?

It is always desirable to cut below a hollow in a limb if possible. Where, however, this would necessitate cutting below the desirable laterals, the cavity may be filled with cement and thus rendered serviceable for some years. Summer pruning of the apricot is desirable if the growth is heavy and the tree has reached a bearing age. Thinning out of prune trees can be undertaken in the autumn, providing the tree has practically finished its growth, as indicated by the change in the color and pose of the leaves.

Apricot Propagation.

Can Royal apricots be grafted into seedling apricots? Do the scions do well? What is the best time to graft them?

The apricot is grafted readily by the ordinary cleft grafting, amputating above the forks if the tree is low-headed enough to allow you to work into the limbs instead of the trunk. Grafts will take all right in the trunk by bark grafting, but working in smaller limbs makes a stronger tree. This is for old trees and the grafting is done during the winter. Younger seedlings can be cleft or whip grafted in the stems, but it is better to bud into the young seedlings with plump buds of the current year's growth, in June, and by shortening in the seedling above the buds as soon as they have taken, get a growth on the bud in the latter half of the same growing season. In nursery practice, trees are usually made by budding in July or August into seedlings which are then growing from the seed planted the previous winter. Little seedlings from under old trees may be carefully transplanted to nursery rows in the spring and budded the same summer. Cultivated well and irrigated if necessary, they will not suffer from this transplanting.

Renewing Old Apricots.

Shall I prune back heavily a 15 to 20-year-old apricot tree which did not mature its fruit this season, I think on account of neglect? It was very poorly cultivated and not irrigated, consequently looks very sick.

Cut back all the main branches to six or eight feet from the ground, leaving on whatever small growth there may be below that height. Paint the stubs and thin out the shoots next summer to get the right number of new branches properly distributed. Whether you will get a good renewal of the head depends upon whether the sickness is in the root or not. Cut back just before the buds swell toward the end of the dormant season.

Summer Pruning of Apricots.

Is it feasible to prune five-year-old apricot trees in August? They seem in good growth and have been irrigated three times this season, though they have never been pruned very closely.

Summer pruning would be perfectly proper and advisable. Summer pruning immediately after the fruit is picked, has become much more general, and winter pruning has proportionately decreased. Young trees are winter pruned to promote low branching and short, stout limbs; bearing trees are summer pruned to promote fruit bearing and check wood growth - the excess of bearing shoots being removed by thinning during the winter.

Wild Cherries.

Where do the Mahaleb and Mazzard cherries grow naturally? How large are the trees, and what kind of fruit do they bear?

The Mazzards, of which there are many, and some of them wild in the Eastern States, are counted inferior seedlings of the species avium, and are tall, large trees, the fruit being small and rather acrid and colors various. The Mahaleb is a European type with a smaller tree, fruit inferior to the Mazzards, and used as a root under soil and climatic conditions under which the Mazzard is not hardy and vigorous. Neither of the kinds are worth considering for their fruit.

Pruning Cherries.

I have some cherry trees that have not been pruned. They are beautiful trees, but it a requires a 24-foot ladder to get near the top limbs. The side limbs reach from tree to tree. They had a splendid crop this year. People here tell me never to prune cherry trees. One man who claims considerable experience with fruit says prune them as soon as the crop is off.

Your cherry trees should have been pruned for the first two or three years quite severely, in order to secure better branching and strength in the main branches. If this is done, and the trees come into full bearing, very little pruning has to be done afterward, except removing diseased, interfering or surplus branches, if there are too many. It is perfectly safe to cut back the trees which you now have as you have been advised to do, after the leaves have fallen or after they have begun to turn yellow. The trees can be safely topped and thinned, for the cherry accepts pruning very readily. Even considerable amounts of the tops have been cut off at fruit-picking time from trees which have been running too high, so that the fruit could be secured, and this has not injured the trees, according to our own experience and observation. Cherries can be summer-pruned to check excessive growth and to promote fruit-bearing, but as your trees have already begun to bear well, this treatment does not seem to be necessary. You should do fall and winter pruning for the shape of the trees.

Training Cherry Grafts.

I have grafted a lot of seedling cherries, leaving two or three buds on each piece of grafted wood. In planting these out, shall I put the union under ground (they are grafted at the crown of the root) and shall I loosen the cloth a little later when they start to grow? How can I get the head for the tree? Should I let only one shoot form, and when it is as high as I want it, cut it off as I would a tree gotten from a nursery?

If you have used waxed cloth in your grafting, it will be necessary to loosen it after the tree gets a good start. Common unwaxed cloth could be trusted to decay soon enough, probably, but it should be looked at to see that it is not binding. The union should not be placed much below the ground surface, although it can be safely covered, and the future stem may look the better for it. One shoot could be allowed to grow from each graft, choosing the best ones and pinching the others so that they will stop extension and hold leaves during the first season. These can be cleanly removed at the first winter pruning at the time you head back the main shoot to the proper height.

Restoring Cherry Trees.

I have about two acres of cherry trees in Sonoma county said to be about 20 years old. They are in a very neglected condition and I am desirous of putting them in good shape for next year's crop. They are in a very light sandy loam sail which is easily worked.

Cherry trees under good growing conditions and proper care are very long lived in California and bear abundant crops when thirty and more years of age. In the San Jose district and elsewhere there are orchards considerably older than the limit stated and are still very profitable. If your trees have been so neglected that the branches have died back, the trees should be pruned, of course, cutting out all dead wood and shortening weak or dying branches to a point where a good strong shoot can be found. Then a good application of farmyard manure plowed in during the rainy season, followed by summer cultivation for moisture retention. Although the cherry is very hardy, it is quite likely to suffer on light soils which become too dry. On such soils as yours there is little if any danger of too much water in the winter, unless the land lies low, but the injury to the tree comes from the lack of moisture during the summer time, and this, with your abundant rainfall, you can probably assure by thorough summer cultivation.

Renewing Cherry Trees.

We have cherry trees set out diamond shape about 16 feet apart. We cannot take out every other tree and have any order, so we ask you if it would be possible to cut the trees back and keep them pruned down to a smaller size. The trees are about 20 years old and are dying back quite badly.

If the trees are dying for lack of summer moisture it is idle to do much for them until you can give them irrigation right after the fruit ripens. The cherry tree takes kindly to cutting back and will give good new fruit-bearing shoots if the roots are in good condition. It is desirable to remove surplus branches entirely rather than to cut back everything to a definite height, the branches to be removed being those which show disposition to die back and those which are running out too far so as to reduce the space between the trees or to interfere with branches from other trees. Branches which are failing above can in some cases be cut back to a strong thrifty lateral branch below. Shortening-in branches high up is less desirable because it forces out too much new growth in the top of the tree and carries the fruit so high that picking would be expensive. All cuts of any size should be painted to prevent the wood from checking.

Pruning Cherries.

I have cherry trees in their third season which have been given the usual winter pruning. The trees are putting forth a great many more branches than are required, and naturally many of the branches are growing across the tree. In cutting these extra branches, I am informed that there is a way to trim them so that they will eventually form fruit spurs. I had an idea that in order to do this it would be well to cut about one inch from the main branch. Some one has told me that this would merely cause the little branch to sprout again.

Cherry shoots which are not required or desired for branch-forming can be transferred into fruit spurs, if the tree is of bearing age, by shortening them in. Do not, however, cut at an arbitrary distance of one inch from the starting point, but rather save one or two buds at whatever distance from the starting point these may be growing. If the tree is too young to bear, only growth shoots may appear from these buds, but they are likely to be short and will support fruit spurs later. This practice should not be carried to excess or you will have too many small shoots which will not get light enough to bear good fruit, even if fruit spurs should appear.

Pollination of Black Tartarian.

There are many old Tartarian cherry trees around our district that have only borne a few cherries in years. There are Bing, Royal Ann and Early Purple Guignes here with these, but they seldom, if ever, bloom with the Tartarian at the proper time to pollinate. What varieties would cause the trees to bear?

Sterility of the Black Tartarian is rather unusual. In the coast regions, Bing, Black Tartarian and Early Purple Guigne are all considered pollinizers for the Royal Ann. Inversely all these should be pollinizers for the Black Tartarian, if that variety requires such assistance, which we have all along supposed that it did not.

Treatment of Fig Suckers.

A few young fig trees are not growing from the tops, but are sending out suckers, in some cases above and others below the point of grafting. Had I better let these suckers grow and see what comes from them or plant new trees?

Graft near the ground all those which are sending suckers from below the graft. Suckers from above grafting point can be trained into trees by selecting the best, tying to stakes to straighten up and removing all other suckers but the one selected.

No Gopher-proof Fig Roots.

Is it necessary that figs should be grafted in some other roots to keep the gophers from destroying the trees? What root should I order?

Figs are not grown on any other than fig roots and are generally propagated by rooted cuttings for the purpose of avoiding the expense of grafting. The fruit must then be protected by killing the gophers rather than by an effort to get the tree upon a gopher-proof root.

Pollination of Bartletts.

Would Clapp's Favorite be a good pollinizer for the Bartlett as well as the White Doyenne?

The white Doyenne and the Clapp's Favorite usually begin to bloom three or four days later than the Bartlett, but the Bartlett period extends about ten days into the blooming period of the others. Therefore, your question is to be answered in the affirmative; that is, if the Bartlett needs pollination, it will be likley to get it from either of these varieties.

Comice Pears.

Would you plant Comice pears instead of Bartletts, and why? What is their behavior as to bearing? Do they require any different treatment than Bartletts? What roots? Do they need other varieties for pollinizing?

Do not plant Cornice instead of Bartletts except for those who have tested out the Cornice to their production and selling. Though satisfactory in some places, it makes no such wide record of success as the Bartlett and should be planted only on the basis of experience with it. Its propagation and culture are the same as other pears. It takes to the quince all right if you want dwarf trees. We have no record of its pollination needs, but as the Bartlett in California defies its Eastern reputation for self-sterility, it is likely that Cornice may also take care of itself, for it is not handicapped by such Eastern condemnation.

No Pears on Peach.

I saw, the other day, some Bartlett pear grafts in Salway peach trees, and the party informed me that he had seen three-year-old grafts that had pears last season. I would like your opinion, as I always thought that such a union was not possible.

Our opinion is like yours, and seeing some pear grafts set in peach branches would not convince us that they would grow or bear fruit.

Pigs in the Orchard.

I have an orchard of Bartlett pears about fifteen years old, located on sediment land. I desire to set this to alfalfa, and to feed the alfalfa by letting hogs eat it off, thereby leaving the droppings on the land. What I wish to know is this: Will this crop be beneficial or injurious to the trees?

Alfalfa can be successfully grown in an orchard, providing you have irrigation water so that the alfalfa shall not rob the trees of moisture; otherwise it is a very dangerous practice. The practice of running animals of any kind in an orchard is to be condemned. Pigs are particularly liable to injure trees by gnawing the bark, and we have seen fig trees barked clean as high as a pig could reach by standing on his hind legs. Of course, if you try an experiment for your own satisfaction, you will have to watch the pigs very carefully. It is true that growing pasture crops in an orchard and grazing, it off is injurious to trees, because the land lacks proper aeration, and good orchard cultivation is even more necessary in this State than in humid climates. Therefore, unless you are sure of a good water supply for irrigation, it would be altogether safer to give the whole land to the trees and keep them cultivated well, or else dig out the trees and use the land for other purposes.

Dwarf Pears Not Commercially Grown.

Will you kindly give the experience of pear growers in California who have grown the dwarfs? If you can give me the data or refer me to persons who can give data showing that the growing of dwarf pears can be made a commercial success the information will be of great value.

There is no commercial growing of dwarf pears in this State, except some trees owned by the A. Block Company, Santa Clara. The late Mr. Block had an old orchard of dwarf trees, planted perhaps forty or fifty years ago, which he converted into an approach to a standard orchard by removing alternate rows, and the trees being otherwise treated like standards have been satisfactorily producing pears for many years. How far these trees are still on the dwarf roots and how far they have supplied themselves with roots from the variety growth above, we do not know. There is no disposition whatever to plant dwarf trees in this State except among a few amateurs who are making home fruit gardens. In view of the successful growth of standard trees in this State, there seem to be no adequate reasons for recourse to dwarf trees.

Yield in Drying Pears.

What is the loss of weight in drying Bartlett pears?

They run from 7 to 8 lbs. of fresh pears to 1 lb. hard dried. There is quite wide variation according to condition of the fruit. Probably about 7 1/2 to 1 would be as near a realizable ratio as you could get by arbitrary estimate.

Pear Problems.

Kindly let me know the advisability of grafting Bartlett pears onto apple trees. In replanting pears in young orchard, how would it do to take rooted pear suckers, graft the Bartlett on them, and save the cost of nursery stock? Last year my five-year-old Bartlett orchard was full of blossoms, but, though many pears became as large as white beans, the majority of them dropped.

The pear and apple do not make a good union. The grafts may grow for a while, but finally fail. Do not use suckers as stocks. You can dig up some year roots and use them as starters by making root-grafts with Bartlett scions and do better than with suckers, but a good pear seedling is the proper thing either for budding or root grafting. Unless you have some experience in such work, it will be cheaper in the end to buy good nursery trees. The nonbearing of your young trees is probably due to their youth and vigor.

Bees and Pear Blight.

A few years ago, I planted alfalfa between my pear trees and the trees bore a very heavy crop that year. Then blight made its appearance, and it was claimed that the bees carried the blight. I therefore plowed under the alfalfa and destroyed what few beehives I had. If the theory that the bees carry the blight from tree to tree is not correct, I will experiment with alfalfa again this year.

It is true that bees carry pear blight. It is also true that you are not likely to get many pears without bees to pollinate the blossoms. You cannot escape the carriage of the pear blight by removing tame bees, because wild bees are abundant in all parts of the State. The way to overcome the blight is to pursue it by amputation of diseased branches continually, so that there may be no contamination for the bees to carry. You are certainly warranted in continuing your alfalfa growing without regard to this question, using water enough to keep the alfalfa growing well without saturating the soil to the injury of the trees or inducing too much summer growth on them.

Forage Under Sprayed Trees.

Is it safe to use arsenical sprays in a pear orchard in which alfalfa is raised between the trees and afterward cut and fed to cattle?

It was fully demonstrated by experiment about 25 years ago that herbage under trees sprayed with paris green at the rate of 1 pound to 160 gallons of water was not injurious to animals pasturing upon it. We are not aware that such an experiment has been made with the more recently used arsenates - which can be used with a much higher amount of arsenic to the gallon because they do not injure the foliage - to determine whether the herbage below would be poisonous or not. Presumably not, because modern spraying does not admit as much loss from run-off as was the case with old Spraying methods.

Pears on Quince.

I saw some time ago a report of some French experiments in grafting the pear onto quince root. The report said the fruit produced was much larger than on any other root.

Most of our common pears will take readily when grafted on the quince, but the quince transforms them into dwarfed trees. Such trees do produce, with proper care, very fine fruit. The remark about their being better than on standard trees refers, however, to other climates than ours, for California grows just as large pears on standard trees as can possibly be grown, while where conditions are harder the higher culture of the dwarf tree and the protection which it requires from climatic hardships, gives the dwarf tree the advantage. You can get pears on quince roots from most of our California nurseries.

Pollination of Pears.

Is it necessary in growing the Comice pear successfully, to put some other pear near for the purpose of pollination in order to make it successful? Will the ordinary Bartlett pear do for pollination?

The Comice pear blooms with the Bartlett, and would therefore presumably be of pollinizing benefit to the Bartlett if the latter should require such treatment. Common experience in California, however, is that the Bartlett is self-fertile and not self-sterile as it is commonly reported in Eastern publications. California practice is, then, to plant Bartletts solidly without reference to preparation for pollination. Taking the matter the other way around, the Bartlett will do for pollination of the Comice probably, if that should be necessary.

Lye-Peeling Peaches.

Please give the formula for peeling peaches by dipping them in caustic soda or lye.

Lye for peeling peaches is used at the rate of half to one pound to the gallon of water, according to the strength of the lye, which you can determine by the quickness with which it acts. The lye water is kept boiling, and the fruit is dipped in wire baskets, only being allowed to remain in the lye a few seconds, and is then plunged at once into fresh water. You must be careful to keep the lye boiling hot, also either to use running water for rinsing or change it very frequently, for you have to rely on fresh water to remove the lye, or the fruit is likely to be stained.

Aged Peach Trees.

What should be done with peach trees 35 years old which are becoming unthrifty, bearing only at the ends of the limbs, etc.?

Old peach trees become bark-bound and need to be cut back to just above the crotch for the forcing out of new branches, this being facilitated, of course, by application of manure, good cultivation of the soil, use of water during the dry season, etc. The peach is, under most conditions, not a long-lived tree, and if your trees are 35 years of age, it is probable that best results could be obtained by grubbing them out and replanting with young trees on new soil if possible. The profitable life of the Eastern peach tree is put down at five or six years. In California the profitable life of the peach sometimes reaches twenty or more years, if growing under exceptionally good conditions; but 35 years would seem to be at least on the borders of decrepitude. Growing at the tips shows that you have not pruned annually to induce the growth of new wood lower down.

Renewing Peach Orchard.

Which is the best way to renew an old peach orchard? The trees are about 18 years old, Muirs and Fosters, and are yielding good crops, but some of the trees show decline. Is it best to replace the old ones with new trees or to plant a new orchard in between the old trees and cut out old ones when new trees are three or four years old?

If the trees have sound bodies and are not badly injured by sunburn borers, do none of the things you mention, but would cut back for a new head. Cutting back should be done during the latter half of the dormant period and thinning of shoots to proper balance a new head should be carefully done the following winter. It is a hard job to get young trees to start among old trees and you are apt to get a mixed lot of trees which you will not be proud of. Cut back as suggested or rip out, plow deeply and start anew, placing the rows midway between the old rows.

Will He Have Peaches?

I have a young orchard between five and six years old, mostly of the Lovell variety. I didn't have much of a crop this year. Should I have a good crop next year?

You ought to be able to tell now how full a set of fruit buds you have. If you do not know what the fruit buds are, ask some neighbor who knows peaches to point them out. If you have a good show of fruit buds, the question in California is not whether they will winter-kill or not, but whether the leaves held late enough the preceding summer and therefore the tree had strength enough to make good strong fruit buds. The late action of the leaves shows that the trees had enough autumn moisture. You will soon learn to recognize the condition also from the plumpness of the wood which carries the fruit buds. If all has gone well so far, the next point is to spray with the bordeaux mixture in November or December so that the new wood shall not be attacked by the peach blight or shothole fungus. This disease comes on early in the winter, sets the the new bark to gumming and endangers the crop. Then if you have San Jose scale, or if your trees showed much curl-leaf last spring, you ought to spray before the blossom buds show color with the lime-sulphur wash. Supposing that you have good buds now and are willing to protect them as suggested, your trees may be expected to come through with a good crop if seasonal moisture conditions are right.

Peach Fillers in Apple Orchard.

I have heard some talk against planting peach fillers in an apple orchard. What is your opinion on the subject?

There is no objection providing the peach is profitable in the locality; and that point you must look into. The peach trees will not injure the apples unless they are allowed to stand too long. In that case they would interfere with the development of the apple.

Grafting Peach on Almond.

May I expect to get good results by grafting some kind of peach to 19-year-old almond tree? If so, what kind of peach will be best? When shall I do grafting?

Peaches take to the almond all right. Cut off and graft in the branches above the main forking of the tree; leaving at least one large branch to be grafted later or to be cut out entirely if you have peach growth enough to fill the top sufficiently. Graft in any kind of peach you find to be worth growing. Graft toward the latter part of the dormant season, say when the buds are swelling for a new start.

Peaches on Apricot.

I have a three-year-old peach orchard grafted or budded on apricot roots, and interspersed through the orchard are young apricot trees, from half-inch to inch and a half in diameter, which sprang from the root, the peach bud or graft having died. I budded these over to peaches in summer, but the buds all died for some cause. What is now the best course to transform them into peach trees? If a graft, what form of graft, and approximately when should it be made?

You can graft peach scions into the apricot sprouts by taking the peach scions of the varieties you desire while the tree is perfectly dormant, keeping them in a cool place and putting in the grafts just as the buds are beginning to swell on the apricot stock. The scions can be buried in the earth in the shade of a fence or building, selecting a place, however, which is moist enough and yet where the water does not gather. The ordinary form of top grafting in stems an inch or more in diameter will work well. The half-inch stems can be whip-grafted successfully. You will have to wax well and see that the wax coating is kept sound until the growth starts.

74

Replanting After Root-knots.

In digging out some old peach trees, I find now and then a tree affected with root knot. I am burning the root, of course, but as these trees are scattered in the orchard, I wish to plant young trees in same locations, thus preserving the rows. Can new stock be safely put in the earth from which the old tree is removed? If treatment of the soil is essential, what is recommended?

Dig a good large hole, removing the earth, and fill with new earth from between the rows, and in this way healthy growth ought to be obtained, although there is always a disposition in some trees to put on knots. They should be looked at from time to time and all those affecting the larger stem should be removed and the wound painted with bordeaux mixture.

Buds in Bearing Trees.

In budding over some old peach trees, should I cut away the branch above the bud when the latter seems to have taken?

The sap flow to the upper part of the branch should be checked by part girdling or by part breaking or bending the top above the bud, after the bud is seen to have set or taken. Do not remove the whole top until the growth on the bud has started out well or else you will "drown it" with excessive sap flow.

Pollen Must Be of the Same Kind.

Do peaches, nectarines and apricots set fruit with the pollen of one another, and are the various peaches, nectarines and apricots self-sterile, or will most kinds set fruit with their own pollen?

We do not count upon pollination between different kinds of fruit. Most fruits are self-fertile, else we could not attain the practical results we do, because it is only in the planting of almonds, cherries, pears and apples that any regard is paid to the association of varieties for that cross-fertilization. Some fruits are more apt to be self-fertile in this State than in other States where the growing conditions are not so favorable.

Peach Budding.

Which is easier with the peach, grafting or budding?

The peach is rather a difficult tree to graft, and budding, on the other hand, is quite easy. You can bud into new shoots of this season's growth in July, and, if necessary, you can improve the slipping of the bark by irrigation a few days before budding. Buds can also be successfully placed in June in the old bark of the peach, providing it is not too old. For this select well-matured buds from the larger shoots and use rather a larger shield than in working into new shoots. When the buds are seen to have taken, the top growth beyond it can be reduced gradually and some new growth forced on the buds the same season, if the sap flow continues as it might be expected to do on young trees well cared for.

Grafting on the Peach.

Will pears do to graft on the peach, or will plums do well on the peach? How soon ought they to bear when grafted on the peach which is past three years old?

Pears cannot be grafted on peaches. Plums generally do well on the peach, and if the grafts are taken from bearing trees, should come into fruit the second season. The peach is more difficult to graft than other fruit trees, because of the drying back of the bark. Be extra careful in the waxing and be sure that the waxing remains good until the growth starts out well the following summer.

Young Trees Failing to Start.

Some peach and almond trees set out last spring lived, but made no growth. Should they be replaced with new stock? If not, what may be expected of them?

If your inactive trees have good plump dormant buds (though they may not be large buds), they may make good growth the coming summer, if the land is good and the moisture right for free growth.

Peach Planting in Alfalfa Sod.

Is it advisable to plant canning peaches in April, and will I gain time in growth and development? I want to set out eight acres in Tuscans or Phillips on deep rich soil near Yuba City. I have a pumping plant and can irrigate. The land has been in alfalfa for several years. I have in mind setting out trees without disturbing the alfalfa - until next plowing season. Do you think it advisable to use commercial fertilizer on ten-year-old Muirs?

Planting the best canning peaches on good peach soil near Yuba City seems to be about the safest line of fruit investment which can be undertaken. We doubt that you can get much growth from trees planted in an old stand of alfalfa without some effort to kill out the plant which now occupies the ground. Still, by deep digging, throwing out all the alfalfa roots and thorough hoeing during the growing season and keeping the alfalfa mowers from sawing off the tops of them, the trees may make a good start. As the alfalfa will have to be irrigated, April may not be too late to start the trees, providing you can find nursery stock which is still quite dormant. Probably ten-year-old peach trees will be very much improved by commercial fertilizers.

Prune on Almond.

What root is considered best for prune trees? The ranch lies above the creek. A friend is very partial to the almond root instead of the myrobalan, but I understand that the prune tree sometimes outgrows the almond root.

If you have a deep rather light soil which drains well and which there is, therefore, no danger of water standing during the rainy season, the almond root is perfectly satisfactory for the prune. It is a strong-growing root and keeps pace with the top growth well. The prune, in fact, is more apt to overgrow the myrobalan than the almond, and the myrobalan will not do well on light soils likely to dry out as the almond will.

Re-grafting Silver Prunes.

I have five acres of Silver prunes which produce very little fruit. The trees are strong and healthy. French prune trees adjoining bear regularly and heavily. Can I graft French prunes on the Silver trees? Will Silver prune trees take other grafts, such as apricots or apples?

The Silver prune is often unsatisfactory for reason of shy bearing. It is perfectly feasible to graft over the tree to the French prune and this has been done for years by different growers. Apricots will usually take on the plum stock, but are apt to over-grow it or else be dwarfed themselves, but the apricot is often worked upon a plum stock. Apples have no grafting affinity whatever for the plum.

French or Italian.

In the prune-growing district around Salem, Oregon, Italian prunes are grown exclusively for drying purposes. French prunes were considered worthless. Here in Sutter county, California, a great many French prunes are grown and we are advised to plant them, but would rather plant the Italian prune. Which would you advise us to set out in this part of the State?

The Italian or Fellenberg prune was grown to some extent in California 40 years and abandoned; it was not so sure in bearing as the French, and it was not the type of prune which we had ambition to excel with. The prune which we grow as the French is the true prune or plum of Agen. We should plant it and let the Oregon people have the Italian.

Myrobalan Seedlings.

I am sending two small plums which I am told are Myrobalan plum. I desire to grow seedlings on which later to bud and graft

French prunes. If these are Myrobalan plums, will trees from them be as good as trees from pits that were imported?

The fruits are Myrobalan plums, and their seedlings would be suitable for the French prune, providing the trees which bear them are strong, thrifty growing trees. There is great variation in the colors of the Myrobalan seedlings, from light yellow to dark red, and it is the satisfactory growth of the tree rather than the character of the fruit which one has to bear in mind when growing seedlings from selected trees instead of depending so largely on imported seedlings.

Drying Plums and Prunes.

I have plum trees of various kinds that are loaded with fruit. I do not know if any are of the variety used for drying as prunes: I know nothing of the process of making or drying prunes. One man suggests that I dip them for four or live minutes in a 3 or 4 per cent solution of lye and then place them in the sun.

Dipping your plums is right providing they are very sweet, as they will dry like prunes without removing the pit. If they are plums that are commercially used for shipping, without enough sugar to dry as prunes, the pit must be removed. Drying in this way, you do not need to use lye, which is simply for the purpose of cracking the skin so that the moisture can be more readily evaporated. There is no danger in using the necessary amount of lye. Less is used than in making hominy.

The Sugar Prune.

What is the commercial value of the Sugar prune? Is there any other early ripening variety better than the Sugar?

It is selling very well as a cured prune, and growers in the northern bay counties especially have done so well that they are extending their plantings. It is coarser in flesh than the French and generally flatter in flavor when cooked and thus falls below the ideal of a cured prune, but it has compensating characters, such as early ripening, with which no other prune compares. The Sugar is also valuable as a shipping plum to Eastern markets.

Glossing Dried Prunes.

Will you give the method for giving the gloss to dried French prunes?

There are various methods. One pound of glycerine to 20 gallons of water; a quick dip in the mixture very hot gives a good finish. Where a clear bloom rather than a shine, is desired, five pounds of common salt to 100 gallons of water, also dipped hot, gives a good effect. Some use a thin syrup made by boiling small prunes in water (by stove or steam) and thinning with water to produce the result desired. Steam cooking avoids bad flavor by burning. The salt dip is probably the most widely used.

Price of Prunes on a Size Basis.

Explain the grading in price of prunes. For instance, if the base price is, say, five and three-fourths cents, what size does this refer to, and how is the price for other sizes calculated? Also, what is the meaning of the phrase "four-size basis"?

Prunes, after being sold to the packer, are graded into different sizes, according to the number required to make a pound, and paid for on that basis. The four regular sizes are 60-70s, 70-80s, 80-90s, and 90-100s, which means that from 60 to 70 prunes are required to make a pound, and so on. The basis price is for prunes that weigh 80 to the pound. When the basis price is 5 3/4 cents, 80-90s are worth 1/4 cent less than this amount, or 5 1/2 cents. The next smaller size, 90-100s, are worth 1/2 cent less, or 5 cents, while prunes under this size are little but skin and pit and bring much less to the grower. For each next larger size there is a difference of 1/2 cent in favor of the grower, so that on the 5 3/4-cent basis 70-80s are worth 6 cents, and 60-70s 6 1/2 cents. This advance continues for the larger sizes, 30-40s, 40-50s, etc., but these quite often command a premium besides, which is fixed according to the supplies available and the demand for the various sizes. The sizes for which no premium or penalty is generally fixed are those from 60 to 100, four sizes, so that this basis of making contracts and sales is called the "four-size basis." The advantage that results in having this method of selling prunes can be seen by the fact that on a 5 3/4-cent basis the smallest of the four sizes will bring but 5 cents a pound, while 30-40s would bring, without any premium, 8 1/2 cents, and with 1 cent premium, 9 1/2 cents. This size has this season brought as high as 10 and 11 cents a pound. It may be noted here that no prunes are actually sold at just the basis price, as they are worth either less or more than this as they are smaller or larger than 80 to the pound. No matter what the basis price is, there is a difference of one-half cent between each size and the sizes nearest to it.

Pollinizing Plums.

How many rows of Robe de Sergeant prune trees should be alternated with the French prune (the common dried prune of commerce) to insure perfect fertilization of the blossoms?

The French prune is self-fertile; that is, it does not require the presence of other plum species for pollination of the blossoms. It is the Robe de Sergeant prune which is defective in pollination and which is presumably assisted by proximity to the French prune. If you wish to grow Robe de Sergeant prunes your question of inter-planting would be pertinent, but if you desire only to grow French prunes you need not plant the Robe de Sergeant at all.

Cultivating Olives.

How deep should an olive orchard be plowed? I was told that by plowing deep I would injure my trees, in cutting up small rootlets and fibres which the olive extends through the surface soil. Is this so or not?

Plowing olives is like plowing other trees, the purpose being to get a workable soil deep enough to stand five or six inches of summer cultivation, usually. If you have old trees which have never been deeply plowed, you would destroy a lot of roots by deep plowing, and you should not start in and rip up all the land at once. You can gradually deepen the plowing, sacrificing fewer roots at a time, without injuring the trees if they are otherwise well circumstanced. Small rootlets and fibres in the surface soil do not count; they are quickly replaced, and if you do not destroy them, the whole surface soil, if moist enough, will be filled with a network of roots which will subsequently make decent working of the soil impossible.

Moving Old Olive Trees.

Would there be anything gained by transplanting old olive trees 6 to 8 inches in diameter over nursery stock? They would have to be shipped from Santa Clara to Butte county and grafted. Would they come into bearing any sooner and be as good trees? Could the large limbs be used to advantage? Would the fact that they are covered with smut cause any trouble?

Old olive trees can be successfully moved a long distance by cutting back, taking up a ball of earth, and possibly a short distance with bare roots if everything is favorable. But do not for a moment think them worth such an outlay for labor, freight and hauling which such a movement as you mention involves. The trees on arrival would probably only be firewood, and if they lived, the time required in getting a good growth and grafting, etc., would perhaps be as great as in bringing a young tree of the right kind to bearing, and the latter would be a better tree in every way. Large limbs can be split and used as cuttings, but the tree would be growth on one side and decay on the other. Use the smaller limbs for hard-wood cuttings and the balance for firewood. The smut shows that the trees are covered with scale insects and might indicate that it is better to burn up the whole outfit unless you learn to fight them.

Darkening Pickled Olives.

Is there anything that will make olives keep their black color when put into lye? When I put my first picking of ripe olives in lye, a large part of them turn green, the black leaving the fruit. My formula is one pound of lye to five gallons of water. Have you any better formula?

By exposing the olives to the light and air, either during the salting or immediately after, ripe olives may be given a uniformly black color. Also, fruit which is less ripe and which shows red and green patches after processing with lye, becomes an almost uniform dark brown color. To do this, the olives are removed from the brine and exposed to light and air freely for one or two days. Your lye was stronger than necessary. With ripe olives it is desirable to use salt and lye together to prevent softening, and the common prescription is two ounces of potash lye and four ounces of salt to the gallon of water after the bitterness is largely removed by using one or two treatments with two ounces of lye to the gallon without the salt. It is necessary to draw off the solution, rinse well, and put on fresh solution several times during the process to get the best results.

Seedling Olives Must Be Grafted.

Will olive trees grown from the olive seed be the right thing to plant?
Will they be true to the parent tree or will they have to be grafted?

Olives which a seedling olive tree will bear will be, as a rule, very inferior and generally of the type of the wild olive. All such trees must be grafted in order to produce any particular variety which you desire.

Olives, Oranges and Peppers.

We have been told that olive trees easily become infested with a fungus disease which they then impart to the orange tree. The same objection is raised to the planting of pepper trees. May this be true in some parts of the State and not in others?

The fungus of which you have heard is the "black smut." It is a result, not a cause. It grows on the honey dew exuded from scale insects and if your trees have no scale they have no fungus. The olive trees and pepper trees may communicate this trouble to citrus trees, or vice versa - whichever gets it first gives it away to the other. If you will work hard enough to kill the scale wherever it appears you can have all these trees, but, of course, it costs a lot to fight scale on big pepper trees, and it is, therefore, wisest usually to choose an ornamental tree not likely to accept the scale.

Budding Olive Seedlings.

I have planted olive seeds which are just sprouting now. Can these be budded next June or July in the nursery row, or can they be bench-grafted the following winter?

Your seedlings may make growth enough to spur-bud this summer. The ordinary plate-bud does not take freely with the olive. Some of them may do this; other seedlings may be slow and have to be budded in the second summer. Watch the size and the sap flow so that the bark will lift well - which may not be at just the time that deciduous trees are budded. It may be both earlier or later in the season. Graft evergreens like the olive in the nursery row; not by bench grafting.

Budding Old Olives.

I have seedling olive trees, set out in 1904, which I wish to change over to the Ascolano variety. Which is the best way to do it, by budding or grafting, and what is the proper time?

Twig-budding brings the sap of the stock to bear upon a young lateral or tip bud, which is much easier to start than dormant buds used either as buds or grafts. A short twig about an inch and a half in length is taken with some of the bark of the small branch from which it starts, and both twig and bark at its base are put in a bark slit like an ordinary shield bud and tied closely with a waxed band, although if the sap is moving freely it would probably do with a string or raffia tie. Put in such buds as growth is starting in the spring.

Olives from Small Cuttings.

In the rooting of small soft-wood olive cuttings is it necessary to cover same with glass - say perhaps prepare a cold-frame and put stable manure in the bottom with about eight inches of sand on top?

It ceases to be a cold-frame when you cover in manure for bottom heat; it becomes a hotbed. Varieties of olives differ greatly in the readiness with which they start from small cuttings. Some start freely and grow well in boxes of sand under partial shade - like a lath house or cover. Some need bottom heat in such a hotbed as you describe with a cloth over; some start well in a cold-frame with a lath cover. To get the best results with all kinds, it is safer to use some more heat than comes from exposure to ordinary temperatures - either by concentration, as in a covered frame, or by a mild bottom heat. If you have glass frames or greenhouse, they are, of course, desirable, but much can be done without that expense.

Olives from Large Cuttings.

I am about to take olive cuttings from one-half to one inch thick and 54 to 20 inches long, and wish to root them in nursery rows. Please advise me if it is necessary to plant under half shade? Also, can same be planted out right away, or should they be buried in trenches for a while before setting out? Would it be best to strip all leaves or branches off, or leave one on? How many buds should be left above ground?

Plant in open ground in the coast district generally; in the interior a lath (or litter shade not too dense) is desirable in places where high dry heat is expected and where sprinkling under the cover may be desirable. Plant out when the soil is right as to warmth and moisture, which is usually a little later than this in the central and northern parts of the State. Remove all leaves and twigs and plant about three-quarters of the length in the soil, which should be a well-drained sandy loam. The cuttings can be taken directly from the trees and need not be bedded. If the cuttings come some distance and get end-dried, make a fresh cut at planting. If shriveled at all, soak a few hours in water before planting out.

Trimming Up Olives.

Limbs are shooting out too low on my olive trees. Would it be right to trim them up while dormant this winter, or should I let them grow another year before doing so? I think I want the first limbs to start at 18 to 20 inches above the ground.

Take off the lower shoots whenever your knife is sharp. Do not let them grow another year. Theoretically, the best time to remove them is toward the end of the dormant season, but if they are not large as compared with the whole growth of the tree, go to it any time.

Canning Olives.

What is the recipe for preserving olives by heat, and how long do they have to remain in the heated state?

Canning olives is a process, not a recipe, and it has to be operated with judgment. It resembles, of course, the common process of canning other fruits and vegetables. It has been demonstrated that heating up to 175° Fahrenheit is effective to keep olives in sealed containers for over two years. The heating was done in the jars in the usual canning way for several minutes after 175° was reached, to be sure the contents were heated through.

Renewing Olive Trees.

I have olive trees on first-class land; no pest of any kind is apparent. The trees look healthy in every way, and average about 12 inches at the butt and 30 feet high. They have borne fruit, but for the last three years have not borne. I am advised to cut back to stumps, 5 or 6 feet high, and start new tops.

Unsatisfactory olive trees may be cut back, but not to such an extent as you mention. Thin out the branches if too thick and cut back or remove those which interfere, but to cut back to a stump would force out a very thick mass of brush which you would have to afterward go into and thin out desperately. The branches which you decide to retain may be cut back to twelve or fifteen feet from the ground. This would have the effect of giving you plenty of new thrifty wood, which is desirable for the fruiting of the olive, but we cannot guarantee that this treatment will make the trees satisfactory bearers. Are you sure they are receiving water enough? If not, give them more next summer. Also give the land a good coat of stable manure and plow under when the land is right for the plow.

Growing Olives from Seed.

How are seedlings grown from olive seeds?

Growing olives from seeds is promoted by assisting nature to break the hard shell. This can be done by pinching carefully with ordinary wire pliers until the shell cracks without injury to the kernel, or the shell may be cut into with a file, making a very small aperture to admit moisture. The French have specially contrived pliers with a stop which admits cracking and prevents crushing. Olive seeds in their natural condition germinate slowly and irregularly. They must be kept moist and planted about an inch deep in sandy loam, covering with chaff or litter to prevent drying of the surface. Before experimenting with olive pits, crack a few to see if they have good plump kernels. Seedling olives must be grafted, of course, to be sure of getting the variety you want. For this reason growth from cuttings is almost universal.

Neglected Olive Trees.

I have a lot of olive trees which have grown up around the old stumps. They are large trees and some of them have six or eight trunks. Should I cut away all but one trunk or let them alone? There are some of the trees with small olives; others none.

If the olive trees which were originally planted were trained at first and still have a good trunk and tree form, the suckers which have intruded from below should be removed. If, however, the trees have been allowed to grow many branches from below, so that there is really no single tree remaining, make a selection of four or five of the best shoots and grow the trees in large bush form, shortening in the higher growth so as to bring the fruit within easier reach and reduce the cost of picking. You can also develop a single shoot into a tree as you suggest. Of course, you must determine whether the trees as they now stand are of a variety which is worth growing. If they are all bearing very small fruit, it would be a question whether they were worth keeping at all, because grafting on the kind of growth which you describe would be unlikely to yield satisfactory tree forms, though you might get a good deal of fruit from them.

Olives from Cuttings.

I have two choice olive trees on my place. I am anxious to get trees from these old ones and do not know how to go about it. Can I grow the young trees by using cuttings or slips from these old trees ? If so, when is the proper time to select the cuttings, and how should they be planted?

Take cuttings of old wood, one-half or three-quarters of an inch in diameter, about ten inches long, and plant them about three-quarters of their length in a sandy loam soil in a row so water can be run alongside as may be necessary to keep the soil moist but not too wet. Such dormant cuttings can be put in when the soil begins to warm up with the spring sunshine. They can be put in the places where you desire them to grow in one or two years. Olives, like other evergreen trees, should be transplanted in the spring when there is heat enough to induce them to take hold at once in their new places, and not during the winter when dormant deciduous trees are best transplanted.

Water and Frost.

I have in mind two pieces of land well adapted to citrus culture. Both have the same elevation, soil, climate and water conditions, except that one piece is a mile of the Kaweah river, while the other is four or five miles distant. In case of a frost, all conditions being about the same, which piece would you consider to be liable to suffer the more? In the heavy frost of last December, while neither sustained any great damage, that portion of the ground nearer the river seemed to sustain the less. Is this correct in theory? The Kaweah river at this point is a good-sized stream of rapidly flowing water.

The land near the river, conditions of elevation being similar, would be less liable to frost. There are a good many instances where the presence of a considerable body of water prevents the lowering of the temperature of the air immediately adjacent. It is so at various points along the Sacramento river, and it is recognized as a general principle that bodies of water exert a warming influence upon their immediate environment even in regions with a hard winter. How much it may count for must be determined by taking other conditions into the account also.

Thinning Oranges.

Is it advisable to thin fruit on young citrus trees? Our trees have been bearing about three years, but they are still small trees. The oranges and grape fruit ripen well and are large and of excellent quality, but the trees seem overloaded.

The size of oranges on over-burdened trees can be increased by thinning, just as other fruits are enlarged, but it is not systematically undertaken as with peaches and apricots, because it is not so necessary and because it is easy to get oranges on young trees too large and to be discounted for over-sized coarse fruit. Removing part of the fruit from young trees is often done - for the good of the tree, not for the good of the fruit. It should be done after the natural drop takes place, during the summer.

Wind-blown Orange Trees.

What would you do for citrus trees five years old that have been badly blown out of shape?

Such trees must be trued up by pruning into the wind; that is, cutting to outside buds on the windward side and to inside buds on the lee side; also reducing the weight by pruning away branches which have been blown too far to the leeward. Sometimes trees can be straightened by moving part of the soil and pulling into the wind and bracing there by a good prop on the leeward side, but that, of course, is not practicable if the trees have attained too much size.

Handling Balled Citrus Trees.

I have some orange and lemon trees which were sent me with their roots balled up with dirt and sacks. As we are still having frosts I have not wanted to set them out. Would it not be better to let them stay as they are and keep the sacks wet (they have a sack box over them) than to put them out while the frosts last?

Your citrus trees will not be injured for a time unless mold should set in from the wet sacks. Get them into the ground as soon as the soil comes into good condition, and cover the top for a time after they are planted to protect them against frosts. This would be better than to hold them too long in the balls, but do not plant in cold, wet soil; hold them longer as they are.

The Navel Not Thornless.

I have lately purchased some Washington navel orange trees, and upon arrival I find they have thorns upon them. I thought the Washington navels were thornless.

The navel orange tree is not thornless. It is described as a medium thorny variety, so that the finding of thorns upon the trees would not be in itself sufficient indication that they were not of the right variety.

Over-size Oranges.

I have some orange trees in a disintegrated granite with a good many small pieces of rock still remaining in the soil. What I wish to know is whether it is probably something in the soil that makes them grow too large, or is it probably the method of treatment? What treatment should be adopted to guard against this excessive growth?

Young trees have a natural disposition to produce outside sizes of fruit, and this is sometimes aggravated by excessive use of fertilizers, sometimes by over-irrigation. We would cease to fertilize for a time and to regulate irrigation so that the trees will have enough to be thrifty without undertaking excessive growth. Such soil as you describe is sometimes very rich at the beginning in available plant food, and fertilization should be delayed until this excess has been appropriated by the tree.

Budding or Grafting in Orange Orchard.

I have land now ready to be planted to oranges, but it is impossible for me to buy the necessary budded stock now or even later this year. Would you advise me to plant the "sour stock" as it comes from the nursery and have it budded or crown-budded later? Are there any real objections to this method, and, if so, what are they?

It is perfectly feasible to plant sour-stock seedlings and to graft them afterward to whatever variety of oranges you desire to grow, but it is undoubtedly better to pay a pretty good price for budded trees of the kind you desire rather than incur the delay and the irregular growth of young trees budded or grafted in the field. There is also danger of an irregular stand from accidental injuries to new growth started in the field without the protection which it finds in the nursery row.

Budding Oranges.

How late in the fall can budding of orange trees be done - plants that are two years old - and what advantage, if any, is late budding? What shall I do with some old trees that were budded about two months ago and are still green but not sprouted yet? The budding was done on young shoots.

Late budding of the orange can be done as late as the bark will slip well; usually, however, not quite so late as this. Such buds are preferred because in the experience of most people they make stronger growth than those put in in the spring. Such buds are not expected to grow until the lowest temperatures of the winter are over. The buds which you speak of as green but still dormant are doing just what they ought to do. They will start when they get ready.

Under-pruning of Orange Trees.

My Washington Navels have a very heavy crop on the lower limbs, as is usual. These branches are so low down that many of the oranges lie on the ground, and it takes a good deal of time to prop them up so that they will not touch the ground. What would be the result of pruning off these low branches, after the fruit is off? Will the same amount of fruit be produced by the fruit growing on the limbs higher up?

Certainly, raise the branches of the orange trees by removing the lowest branches or parts of branches which reach to the ground. A little later others will sag down and this under-pruning will have to be continuous. It would be better to do this than to undertake any radical removal of the lower branches. The progressive removal as becomes necessary will not appreciably reduce the fruiting and will be in many ways desirable.

Keeping Citrus Trees Low.

My tangerines last fall shot up like lemon trees - a dozen to twenty shoots two or three feet high. The trees are eight years old and are loaded with bloom and some of the shoots have buds and bloom clear to the top. Some shoots have no bloom. What should I do with these shoots? Cut them back like lemons or let them remain?

You must shorten the shoots if you desire to have a low tree. This will cause their branching and it will be necessary, therefore, to remove some of the shoots entirely, either now or later, in order that the tree will not become too compact.

Dying Back of Fruit Trees.

I have a few orange and lemon trees that are starting to die. One tree has died on the top. What kind of spray shall I use?

The dying back of a tree at the top indicates that the trouble is in the roots, and it is usually due to standing water in the soil, resulting either from excessive application of water or because the soil is too retentive to distribute an amount of water which might not be excessive on a lighter soil which would allow of its freer movement. Dig down near the tree and see if you have not a muddy subsoil. The same trouble would result if the subsoil is too dry, and that also you can ascertain by digging. If you find moisture ample, and yet not excessive, the injury to the root might be due to the presence of alkali, or to excessive use of fertilizers. The cause of the trouble has to be determined by local examination and cannot be prescribed on the basis of a description of the plant. It cannot be cured by spraying unless specific parasite is found which can be killed by it.

Young Trees Dropping Fruit.

I have a few citrus fruit trees about three years old. They have made a good growth and are between seven and eight feet high with a good shaped top or head. I did not expect any fruit last year and did not have any. This spring they blossomed irregularly at blooming time, but quite an amount of fruit set and grew as large as marbles, some of it the size of a walnut, but lately it has about all fallen off the trees.

There is always more or less dropping from fruit trees. Some years large numbers of oranges drop. There may be many causes, and the trouble has thus far not been found preventable. When the foliage is good and the growth satisfactory, the young tree is certainly not in need of anything. It is rather more likely that fruit is dropped by the young trees owing to their excessive vegetative vigor, for it is a general fact that fruit trees which are growing very fast are less certain in fruit-setting. It is, of course, possible that you have been forcing such action by too free use of water. You will do well to let your trees go along so long as they appear thrifty and satisfactory, and expect better fruiting when they become older.

Orange Training.

Is not a single leader in an orange tree more desirable than the much-forked tree so commonly seen! Can a single-leader tree be made from the nursery trees which have already formed their heads, by cutting off the heads below so that only a straight stick without any branches is left?

An orange tree with a central leader would not be at all satisfactory if it were carried very high. Of course, a central stem can be to advantage taken higher than it is often done, but we would not think of growing an orange tree with a central stem to the apex. The laterals would droop, crowd down upon each other badly, open the center to sunburn, and encourage also a growth of central suckers and occasion an amount of pruning altogether beyond what is necessary with a properly branched tree without a central stem.

Curing Citron.

I wish to know a way to cure citrons at home. I have a fine tree that has borne very fine-looking fruit for the past two years.

An outline for the preparation of candied citron is as follows: The fruit, before assuming a yellow color, and also when bright yellow, is picked and placed in barrels filled with brine, and left for at least a month. The brine is renewed several times, and the fruit allowed to remain in it until required for use, often for a period of four or five months. When the citrons are to be candied they are taken from the barrels and boiled in fresh water to soften them. They are then cut into halves, the seed and pulp are removed, and the fruit is again immersed in cold water, soon becoming of a greenish color. After this it is placed in large earthen jars, covered with hot syrup, and allowed to stand about three weeks. During this time the strength of the syrup is gradually increased. The fruit is then put into boilers with crystallized sugar dissolved in a small quantity of water, and cooked; then allowed to cool, and boiled again until it will take up no more sugar. It is then dried and packed in wooden boxes.

Crops Between Orange Trees.

What crop can I plant between rows of young orange trees to utilize the ground as well as pay a little something?

It depends not alone upon what will grow, but upon what can be profitably sold or used on the place, and unless sure of that, it is usually better not to undertake planting between young trees but rather to cultivate well, irrigate intelligently, and trust for the reward in a better growth and later productiveness of the trees. It is clear, California experience that planting between trees except to things which are demonstrated to be profitable should not be undertaken, and where one does not need immediate returns is, as a rule, undesirable. The growth of a strip of alfalfa, if one is careful not to submerge the trees by over-irrigation, would be the best thing one could undertake for the purpose of improving the soil by increasing the humus content, reducing the amount of reflected heat from a clean surface, and is otherwise desirable wherever moisture is available for it. You could also grow cow peas for the good of the land if not for other profit. You can, of course, grow small fruits and vegetables for home use if you will cultivate well. Common field crops, with scant cultivation, will generally cause you to lose more from the bad condition in which they leave the soil than you can gain from the use or sale of the crop.

Navels and Valencias.

Navel trees are being budded to Valencias in southern California, because of the higher price received for the late-ripening Valencias. Are the orchards in central and northern California being planted in Navels, and is there any difference in soil or climate requirements of Navels and Valencias?

There is no particular difference in the soil requirements of Valencia and Navel oranges. They are both budded on the same root. The desirability of Navel oranges in the upper citrus districts arises from the fact that the policy of those districts at the present time is to produce an early orange. This they could not accomplish by growing the Valencia. The great advantage of the Valencia in southern California, on the other hand, lies in the very fact that it is late and that it can be marketed in midsummer and early autumn when there are no Navels available from anywhere.

Orange Seedlings.

What about planting the seed from St. Michael's oranges or of grapefruit for a seed-bed to be budded to Valencias?

Good plump St. Michael's seeds would be all right if you desire to use sweet seedling stock. Grapefruit seedlings are good and quite widely used, though the general preference is for sour-stock seedlings.

Acres of Oranges to a Man.

In your opinion, is it possible for one man, of average strength, to take perfect care of a twenty-acre citrus orchard? Are the services of a man who takes the entire responsibility of an orchard (citrus) worth more than those of a common ranch hand?

It depends upon the man, upon the age of the trees, upon the kind of soil he has to handle, upon the irrigation arrangements and upon what you mean by "perfect care." If you contract the picking and hauling of fruit, the fumigation and allow extra help when conditions require that something must be done quickly, whatever it may be, a man with good legs and arms, and a good head full of special knowledge to make them go, can handle twenty acres and if he does it right you ought to pay him twice as much as an ordinary ranch hand.

Roots for Orange Trees.

What are the conditions most favorable to orange trees budded upon sour stock; also upon sweet stock and trifoliata?

The sour stock is believed to be more hardy against trying conditions of soil moisture - both excess and deficiency, and diseases incident thereto. The sweet stock is a free growing and satisfactory stock and most of the older orchards are upon this root, but it is held to be less resistant of soil troubles than the sour stock, and therefore propagators are now largely using the latter. The trifoliata has been promoted as more likely to induce dormancy of the top growth during cold weather, because of its own deciduous habit. It has also been advocated as likely to induce earlier maturity in the fruit and thus minister to early marketing. The objection urged against it has been a claimed dwarfing of the tree worked upon it.

Citrus Budding.

I wish to bud some Maltese blood orange trees to pomelos and lemons. Will they make good stock for them, and, if so, is it necessary to cut below the original bud?

It is possible to bud as you propose, and it is not necessary to go back to the old stock. Work in above the forks.

No Citrus Fruits on Lemon Roots.

Would it be any advantage to bud the Washington Navel on grapefruit and lemon roots?

The grapefruit or pomelo is a good root for the orange, and some propagators prefer it. The lemon root is not used at present, because of its effect in causing a coarse growth of tree and fruit and because it is more subject to disease than the orange root. In fact, we grow nearly all lemons on orange roots.

Budding Oranges.

My first attempt at budding, I cut 20 buds and immediately inserted in stock of Mexican sour orange "Amataca." I left bands on them for ten days at which time about half seemed to have "stuck," but after a few days the bark curled away and the buds dried up and died. I then tried again, but left the bands on for thirteen days and lightly tied strings around below the bud to prevent the bark from curling, and also put grafting wax in the cut and over the bud. These appeared fresh and green at time of taking off the bands, but three weeks later I found them rotted. The grafting wax used was made of beeswax, resin, olive oil and a small amount of lard to soften it. Do you think that the action of the lard on the buds would cause them to rot?

Consider first whether the buds which you use are sufficiently developed; that is, a sufficient amount of hardness and maturity attained by the twig from which you took these buds. Second, use a waxed band, drawing it quite tightly around the bark, above and below the bud, covering the bud itself without too much pressure for several days, then loosening the band somewhat, but carefully replacing over all but the bud point. It is necessary to exclude the air sufficiently, but not wholly. The use of a soft fat like olive oil or lard is not desirable. If you use oil at all for the purpose of softening, linseed oil, as used by painters, is safer because of its disposition to dry without so much penetration. Having used olive oil and lard together you had too much soft fatty material.

Budding Orange Seedlings in the Orchard.

What are the objections or advantages of planting sour stock seedlings where one wishes the trees and one or two years later bud into the branches instead of budding the young stock low on the trunk?

Planting the seedling and at some future time cutting back the branches and grafting in the head above the forks is an expensive operation and loses time in getting fruit. You will get very irregular trees and be disappointed in the amount of re-working you will have to do. Suckers must be always watched for; that has to be done anyway, but a sucker from a wild stock is worse in effects if you happen to overlook it. Avoid all such trouble by planting good clean trees budded in nursery rows. You may have to do rebudding later, if you want to change varieties, and that is trouble enough. Do not rush at the beginning into all the difficulties there are.

Grapefruit and Nuts.

Peaches, pears and plums predominate in this section, but would not grapefruit, almonds and English walnuts be just as profitable? What is your idea about English walnuts on black walnut root?

You can expect grapefruit to succeed under conditions which favor the orange. Therefore, if oranges are doing well in your district, grapefruit might also be expected to succeed on the same soils and with the same treatment. Planting of almonds should proceed upon a demonstration that the immediate location is suited to almonds, because they are very early to start and very subject to spring frost and should not be planted unless you can find bearing trees which have demonstrated their acceptance of the situation by regular and profitable crops. English walnuts are less subject to frosts because they start much later in the season. They need, however, deep, rich land which will be sure not to dry out during the summer. English walnuts are a perfect success upon the California black walnut root.

Soil and Situation for Oranges.

Is it absolutely essential that orange trees be planted on a southern slope, or will they thrive as well on any slope? What is the minimum depth of soil required for orange trees? How can I tell whether the soil is good for oranges?

Orange trees are grown successfully on all slopes, although in particular localities certain exposures may be decidedly best, as must be learned by local observation. How shallow a soil will suit orange trees depends upon how water and fertilizer are applied; on a shallow soil more fertilizer and more frequent use of water in smaller quantities. Any soil which has grown good grain crops may be used for orange growing if the moisture supply is never too scant and any excess is currently disposed of by good drainage. There can be no arbitrary rule either for exposure, depth or texture of soils, because oranges are being successfully grown on medium loam to heavy clay loam, providing the moisture supply is kept right.

Transplanting Orange Trees.

Can you transplant trees two years old with safety to another location in same grove, same soil; etc.?

Yes; and you can move them a greater distance, if you like. Take up the trees with a good ball of earth, transplanting in the spring when the ground has become well warmed, just about at the time when new growth begins to appear on the tree. The top of the tree should he cut back somewhat and the leaves should be removed if they show a disposition to wilt. You should also whitewash or otherwise protect the bark from sunburn if the foliage should be removed.

Protecting Young Citrus Trees.

Is it necessary to have young orange trees covered or leave them uncovered during the winter months?

It is desirable to cover with burlaps or bale with cornstalks, straw or some other coarse litter, all young trees which are being planted in untried places; and even where old trees are safe, young trees which go into the frost period with new growth of immature wood should be thus protected. Do not use too much stuff nor bundle too tightly.

Not Orange on the Osage.

Can the Navel orange be grafted on the osage orange? I understand it is done in Florida, and would like to know if it has been tried in California.

It cannot. It has not been done in Florida nor anywhere else. The osage orange is not an orange at all. The tree is not a member of the citrus family.

No Pollenizer for Navels.

I read that the flowers of the Navel orange are entirely lacking in pollen, or only poorly supplied. If this is true, what variety of orange would you plant in a Navel grove - to supply pollen at the proper time?

We would not plant any other orange near the Navel for the sake of supplying it with pollen. Pollen is only needed to make seeds, and by the same process to make the fruit set, and Navels do not make seeds, except rarely, nor do they seem to need pollen to make the fruit set.

Water and Frost.

From how many acres could I keep off a freeze of oranges with 1000 gallons per minute? The water is at 65 degrees.

The amount of water will prevent frost over as large an area as you can cover with the water, so as to thoroughly wet the surface, but the presence of water will only be effective through about four degrees of temperature and only for a short time. If, then, the temperature should fall below 27 degrees and should remain at that point for an hour or two, it is doubtful if the water would save your fruit. Water is only of limited value in the prevention of frost, and of no value at all when the temperature falls too low.

What to Do with Frosted Oranges.

What is the best plan of treatment for frosted orange trees? The crop will be a total loss. It does not show any tendency to fall off the trees, however. Should it be picked off, thrown on the ground and plowed under? Should this be done right away or later?

Unsound fruit should be removed as soon as its injury can be conveniently detected and worked into the soil by cultivation; never, however, being allowed to collect in masses, which is productive of decay and which may be injurious to roots. If trees are injured sufficiently to lose most of their leaves, the fruit should also be removed if it shows a disposition to hang on. This will be a contribution to the strength of the tree and its ability to clothe itself with new foliage.

Pruning Frosted Citrus Trees.

How shall I prune two-year-old orange orchard, also nursery stock buds that are badly injured by frost; how much to prune and at what time?

As soon as you can see how far injury has gone down the branch or stem, cut below it, so that a new shoot may push out from sound wood, and heal the cut as soon as possible. This applies to growths of all ages. In the case of buds, if you can only save a single node you may get a bud started there and make a tree of that. In the case of trees, large or small, it is always desirable to cut above the forkings of the main branches, if possible, and when this much of the tree remains sound, a new tree can be formed very quickly. If the main stem is injured, bark cracked, etc., cut below the ground and put scions in the bark without splitting the root crown; wax well or otherwise cover exposed wood to prevent checking. If this is successfully done, root-rot may be prevented and the wound covered with new bark while the strong new stems are developing above.

Pruning Oranges.

Is it best to prune out orange trees by removing occasional branches so as to permit free air passage through the trees? Some are advocating doing so; but as I remember, the trees in southern California are allowed to grow quite dense, so that we could see into the foliage but very little.

It is a matter of judgment, with a present tendency toward a more open tree than was formerly prescribed. Trees should be more thrifty and should bear more fruit deeper in the foliage-wall if more air and light are admitted. But this can be had without opening the tree so that free sight of its interior is possible. We believe thinning of the growth to admit more light and air is good, but we should not intentionally cut enough to make holes in the tree.

Pecan Growing.

Would you advise planting of pecans in commercial orchards here? Walnuts in their proper location constitute some of California's best improvements. After visiting some bearing paper-shell pecans here in Fresno county, I believe a pecan orchard of choice variety would be more desirable than a walnut orchard.

Pecans do well on moist rich land in the interior valleys where there are sharper temperature changes than in the coast valleys, except perhaps near the upper coast. Such planting as you propose seems promising on lands having moisture enough to carry the nuts to full ripening.

Growing Filberts.

Please give information about growing filberts.

Filberts have been largely a disappointment in California and no product of any amount has ever been made. Good nuts have been produced in the foothills of the Sierra Nevada and the Coast Range. Theoretically, the places where the wild hazel grows would best suit the filbert, and so far this seems to be justified by the little that has actually been done, but there is very little to say about it beyond that. It requires much more experience to lift the nut out of the experimental state.

Early Bearing of Walnuts.

Please inform me if young walnut trees grafted on black walnut stock will produce fruit within 18 months after being planted.

It is true that the French varieties of English walnuts have produced fruit the second summer of their growth. This does not mean, however, that you can count upon a crop the second year. These are usually grafts in nursery rows, and one would have to wait longer, as a rule, for trees planted out in orchards with a chance to make a freer wood growth. This is rather fortunate, because it is better to have a larger tree than to have the growth diverted into bearing a small amount of fruit while the tree is very young. We do not know any advantage in getting nuts the second year except it be to see if you really have secured the variety you desire to produce later.

Handling Walnut Seedlings.

What is the best time to transplant seedlings of the black walnut?

Transplant during the dormant season (as shown by absence of leaves) when the soil is in good condition. Handle them just as you would an apple tree, for instance.

How to Start English Walnuts.

In starting English walnuts, shall we get nursery stock grafted on California black, or shall we start our black walnut seedlings in nursery plats, or plant the nuts where the tree is wanted, and graft them at two or three years? What is the advantage, if any, of the long stock from grafting high, over the grafted root?

If we had the money to invest and were sure of the soil conditions, etc., we should buy grafted trees of the variety we desired, just as we would of any other kind of fruit. If we were shy of money and long on time, we would start seedlings in nursery, plant out seedlings, and graft later, because it is easier to graft when the seedling is two or three years in place. We count the planting of nuts in place troublesome and of no compensating advantage. The chief advantage known to us of grafting high and getting a black walnut trunk is the hardier bark of the black walnut.

Walnut Planting.

I am planning to plant walnuts on rather heavy soil. I have been told to put the nut six inches below the surface, but think that too deep, as soil is rather heavy.

In a heavy soil we should not plant these nuts more than three inches below the surface, but should cover the surface with a mulch of rotten straw to prevent drying out.

Pruning Grafted Walnuts.

Should English walnut trees be pruned? I have along the roadside English walnuts grafted on the California black, and they have grown to very large size and the fruit seems to be mostly on the outside of the trees.

English walnuts are not usually pruned much, though it is often desirable, and of course trees can be improved by removing undesirable branches and especially where too many branches have started from grafts, it is desirable that some be removed. They should be cleanly sawed off and the wound covered with wax or thick paint to prevent the wood from decaying.

Pruning Walnuts.

When is the best time to remove large limbs from walnut trees?

This work with walnuts or other deciduous fruit trees should be done late in the winter, about the time the buds are swelling; never mind the bleeding, it does no harm, and the healing-growth over the wound is more rapid while the sap is pushing.

Grafting Walnuts.

In cleft grafting walnuts is it necessary to use scions with only a leaf bud, or with staminate or pistillate buds? Is cutting the pith of the scion or stock fatal to the tree?

In grafting walnuts it is usual to take shoots bearing wood buds, and not the spurs which carry the fruit blossoms, although a part of the graft containing also a wood bud can be used, retaining the latter. Cutting into the pith of the scion or of the stock is not fatal, but it is avoided because it makes a split or wound which is very hard to heal. For this reason it is better to cut at one side of the pith in the stock, and to cut the scion so that the slope is chiefly in the wood at one side of the pith and not cutting a double wedge in a way to bring the pith in the center.

Grafting Nuts on Oaks.

I have 10 to 15 acres of black oak trees which I wish to graft over to chestnuts. Can grafting be done successfully?

Some success has been secured in grafting the chestnut on the chestnut oak, but not, so far as we have heard, on the black oak. But grafts on the chestnut oak are not permanently thrifty and productive, though they have been reported as growing for some time. The same is true of English walnut grafts on some of the native oaks.

Grafting Walnut Seedlings.

Would it be proper to graft one-year California black walnut seedlings that must also be transplanted?

As the seedlings must be moved, plant in orchard and graft as two or three-year-olds, according to the size which they attain.

Pruning the Walnut.

What is the proper time for pruning the walnut? Is it bad for the tree to prune during the active season? I have recently acquired a long-neglected grove in which many large limbs will have to be removed in order to allow proper methods of cultivation to be practiced, and I am in doubt as to the wisdom of doing this during the rise of sap.

The best time to remove large limbs to secure rapid growth of bark from the sides of the cut, is just at the time the sap is rising. There will be some outflow of sap, but of no particular loss to the tree. As soon as the large wounds have dried sufficiently, the exposed surface should be painted to prevent cracking of the wood.

Eastern or California Black Walnuts?

I am told that the Eastern black walnut is a more suitable root for the low lands in California than the California black. Is this true?

There has been no demonstration that the Eastern black walnut is more suitable to low moist lands than the California black walnut. Our grandest California black walnut trees are situated on low moist lands. Walnut Grove is on the edge of the Sacramento river with immense trees growing almost on the water's edge. Walnut Creek in Contra Costa county is also named from large walnut trees on the creek bank land. We have very few Eastern black walnut trees in California and although they do show appreciation of moist land, they are not in any respect better than the Californian.

Ripening of Walnuts.

I send you two walnuts. I am in doubt if they will mature.

The nuts are well grown, the kernel fully formed in every respect. Whether they will attain perfect maturity must be determined by an observation of the fact and cannot be theoretically predicated. Where trees are in such an ever-growing climate as you seem to have, they must apparently take a suggestion that the time has arrived for maturity from the drying of the soil. The roots should know that it is time for them to stop working so that the foliage may yellow and the nuts mature. It is possible that stopping cultivation a little earlier in the season may be necessary to accomplish this purpose.

Cutting Below Dead Wood.

I have some seedling English walnut trees which are two years old, but they are not coming out in bud this year. They are about three feet high, and from the top down to about 10 inches of the ground the limbs are dark brown, and below that they are a nice green. I cut the top off of one of them to see what is the matter that they do not leaf out, and I found that there is a round hole right down through the center of the tree down to the green part. The hole is about three-sixteenths of an inch in diameter. The pith of the limbs has been eaten away by some kind of a worm from the inside.

Would it be better to cut the tree down to the green part, or let them alone?

It is the work of a borer. Cut down to live wood and paint over the wound or wax it. Protect the pith until the bark grows over it or you will have decay inside. If buds do not start on the trunk, take a sucker from below to make a tree of. You could put a bud in the trunk, but it is not very easy to do it.

Walnuts in Alfalfa.

Will the walnut trees be injured in any way by irrigating them at the same time and manner as the alfalfa - that is, by flooding the land between the checks? Will the walnuts make as good a growth when planted in the alfalfa, and the ground cultivated two or three feet around the tree, as though the alfalfa was entirely removed? Is it advisable to plant the trees on the checks rather than between the checks?

Walnut trees will do well, providing you do not irrigate the alfalfa sufficiently to waterlog the trees; providing also that you do use water enough so that the trees will not be robbed of moisture by the alfalfa. This method of growing trees will be, of course, safer and probably more satisfactory if your soil is deep and loamy, as it should be to get the best results with both alfalfa and walnuts. It would be better to have the trees stand so that the water does not come into direct contact with the bark, although walnut trees are irrigated by surrounding them with check levees. Planting walnut trees in an old stand of alfalfa is harder on the tree than to start alfalfa after the trees have taken hold, because the alfalfa roots like to hang on to their advantage. In planting in an old field, we should plow strips, say, five feet wide and keep it cultivated rather than to try to start the trees in pot-holes, although with extra care they might go that way.

Walnuts in the Hills.

Will walnuts grow well in the foothill country; elevation about 600 feet, soil rich, does not crack in summer and seems to have small stones in it?

Walnuts will do well providing the soil or subsoil is retentive enough. If you have water available for irrigation in case the trees should need it, they would do well, but if the soil is gravelly way down and likely to dry out deeply and you have no water available an opposite result might be expected. It is a fact that on some of the uplands of the coast mountains there is a lack of moisture late in the season which interferes with the success of some fruit trees.

To Increase Bearing of Walnuts.

We have a walnut orchard which does not bear enough nuts. The trees are all fine, even trees, 10 and 12 years old, and we are told that the crop was light this year because the trees were growing so vigorously and put most of their energy into the new wood. Is there any special fertilizer which will make the trees bear more and not prompt such heavy growth?

If your adviser is right that the trees are not bearing because of excessive growth, it would be better not to apply any fertilizer during the coming year, but allow the trees to assume more steady habit and possibly even to encourage them to do so by using less cultivation and water. If you wish to experiment with some of the trees, give them an application of five pounds of superphosphate and two pounds of potash to each tree, properly distributed over the land which it occupies. You certainly should not use any form of nitrogen.

Temperature and Moisture for the English Walnut.

What amount of freezing and drouth can English walnuts stand? Under what conditions is irrigation necessary?

The walnut tree will endure hard freezing, providing it comes when the tree is dormant, because they are successfully grown in some parts of the Eastern States, though not to a large extent; but the walnut tree is subject to injury from lighter frosts, providing they follow temperatures which have induced activity in the tree. On the Pacific Coast the walnut is successfully grown as far north as the State of Washington, but even in California there are elevations where frosts are likely to occur when the tree is active, and these may be destructive to its profit, although they may not injure the tree. You are not safe in planting walnuts to any extent except in places where you can find trees bearing satisfactorily. Planting elsewhere is, of course, an enterprising experimental thing to do, but very risky as a line of investment. Irrigation is required if the annual rainfall, coupled with the retentiveness of the soil and good cultivation, do not give moisture enough to carry the tree well into the autumn, maintaining activity in the leaves some little time after the fruit is gathered.

Walnuts from Seed.

There is a reliable nursery company selling seedling Franquette walnut trees on a positive guarantee that they will come true to type. Are orchards of this kind satisfactory?

Walnuts do come truer to the seed than almonds and other fruits and the Franquette has a good reputation for remembering its ancestry. Until recently practically all the commercial walnut product of California was grown on seedling trees. But these facts hardly justify one in trusting to seedlings in plantings now made. The way to get a walnut of the highest type is to take a bud or graft from a tree which is bearing that type.

High-grafted Walnuts.

What is the advantage of a high-grafted walnut? I am about ready to plant 10 acres to nuts and do not know whether to purchase Franquette grafted high on California Black or not.

The advantage of grafting English walnut high on California Black walnut consists in securing a main trunk for the tree, which is less liable to sunburn and probably hardier otherwise than is the stem of the English walnut, and the present disposition toward higher grafting or budding seems therefore justified and desirable.

Grafting and Budding the Mulberry.

What is the most approved manner of grafting mulberry trees? Am told that they are very difficult to successfully graft.

Most propagators find the mulberry difficult by ordinary top and cleft grafting methods. A flute or ring graft or bud does well on small seedlings - that is, removing a ring or cylinder of the bark from the stock and putting in its place a cylinder from the variety desired, cut to fit accurately. For large trees this would have to be done on young shoots forced out by cutting back the main branches, but when this is done ordinary shield budding in these new shoots would give good results. Cut back the trees now and bud in the new shoots in July or August.

118

Hardiness of Hybrid Berries.

How much cold will Phenomenal, Himalaya and Mammoth blackberries stand in winter? Is it safe to plant where the temperature goes below 32 degrees?

These berries are hardy to zero at least, for they are grown in northern parts of this coast where they get such a touch once in a while. They have also endured low temperatures in the central continental plateau States and eastward. Whether they can endure the lowest temperatures of the winter-killing regions of the northern border cannot be determined in California, for we do not have the conditions for such tests. The berries are very hardy while dormant, and probably their value in colder regions would depend rather more upon their disposition to remain dormant than upon what they can endure when in that condition.

Pruning Himalayas.

Shall the old wood be cut away in pruning Himalayas?

All the old wood which has borne fruit should be cut out in the fall and new shoots reduced to three or four from each root, and these three or four shoots should be shortened to a length of ten or twelve feet and be trained to a trellis or fence, or some other suitable support. Vines which are allowed to grow riotously as they will, are apt to be deficient in fruit bearing.

Strawberries with Perfect Flowers.

Has Longworth Prolific an imperfect bloom? I have Longworths in bearing which apparently are perfect. Is there another strain of Longworth that are not self-fertilizing?

The Longworth Prolific strawberry has both staminate and pistillate elements. Possibly some other variety, because of its resemblance to Longworth and the popularity of it, may have been wrongly given its name. Most of the varieties which are largely grown in California are perfect in blossom, though some of the newer varieties need association with pollinizers.

Pruning Loganberries.

Should the new shoots of Loganberry vines, which come out in the spring, be left or cut away? If cut, will more shoots put out in the fall and be sufficient for the next year's crop?

The Loganberry shoots which are growing should be carefully trained and preserved for next year's fruiting. The old canes should be cut away at the base after the fruit is gathered. The plant bears each year upon the wood which grew the previous summer.

Strawberry Planting.

Should I plant strawberries in the spring or fall?

Whether it is wise to plant strawberry plants in the fall depends on several things, such as getting the ground in the very best of condition, abundance of water at all times, splendidly rooted plants, and cool weather (which is very rare at the time plants are to be planted, August and September). Plants may be taken with balls of earth around the roots, and water poured in the hole that receives the plant. After planting, each plant should be shaded from the sun; after this the ditches must be kept full of water so the moisture will rise to the surface; this must be done till the plant starts growth. This method can only be used in small plantings, as it is too expensive for large plantings, as is also the potted plant method where each plant is grown in a small pot and transplanted by dumping out the earth as a ball with the plant and putting directly in the ground. From potted plants, set out in the fall, one may count on a fine crop of berries the following spring. Strawberry plants are never dormant till midwinter, and there is no plant more difficult to transplant when roots are disturbed in the hot season, which usually prevails in the interior valleys of California. To have a long-lived strawberry field and to get best results, planting must be done in the spring, as soon as the soil can be put in best condition to receive plants. From this a fall crop can be expected - Answer by Tribble Bros., Elk Grove.

Blackberries for Drying Only.

What variety of blackberries or raspberries are the best for drying purposes? Are berries successfully dried in evaporators? This is a natural berry country. Wild blackberries are a wonder here. Transportation facilities do not allow raising for the city market. In your opinion, would the planting of ten acres in berries for drying be a success?

The blackberries chiefly grown in California are the Lawton, Crandall and the Mammoth. The raspberry chiefly grown is the Cuthbert. There are very few of these berries dried. It would be better to dry them in an evaporator than in the sun, but little of it is done in this State. It is doubtful whether it would pay to plant blackberries for drying only, because there is such a large product flow in various places where the berries are either sold fresh or sold to the cannery, and drying is only done for the purpose of saving the crop if the prices for the other uses are not satisfactory. To grow especially for drying would give you only one chance of selling to advantage, and that the poorest.

Planting Bush Fruits.

What is the best time to set out blackberries and Loganberries?

Any time after the soil is thoroughly wet down and you can get good, mature and dormant plants for transplanting. This may be as early as November and may continue until February or later in some places.

Growing Strawberry Plants.

In a patch of strawberries planted this spring, is it advisable to cut off runners or root some of them?

In planting strawberries in matted rows, it is usual to allow a few runners to take root and thus fill the row. It is the judgment of plant growers that plants for sale should not be produced in this way, but should be grown from plants specially kept for that purpose.

Strawberries in Succession.

Is there any reason, in strawberry culture, when the vines are removed at the end of the fourth year, why the ground may not be thoroughly plowed and again planted to strawberries?

122

It is theoretically possible to grow strawberries continuously on the same land by proper fertilization and irrigation. Practically, the objection is that certain diseases and injurious insects may multiply in the land, and this is the chief reason why new plantations are put on new land and the old land used for a time for beans or some root crop, so that the soil may be cleaned and refreshed by rotation and by the possibility of deeper tillage.

Limitations on Gooseberries.

Why is it that gooseberries are not grown more in California? Is there any reason, climatic or other, why the gooseberry should not be as successfully grown in California as elsewhere?

There are two reasons. First, the gooseberry does not like interior valleys, although with proper protection from mildew or by growing resistant varieties, good fruit can be had in coast or mountain valleys. Second, practically no one cares for a ripe gooseberry in a country where so many other fruits are grown, and the demand is for green gooseberries for pies and sauce, and that is very easily oversupplied.

Dry Farming with Grapes.

I have heard that they are planting Muscat grapes on the dry farming plan. Will it be successful?

Grapes have been grown in California on the dry farming plan ever since Americans came 60 years ago. Grapes can be successfully grown by thorough cultivation for moisture retention, providing the rainfall is sufficient to carry the plant when it is conserved by the most thorough and frequent cultivation. Unless this rainfall is adequate, no amount of cultivation will make grape vines succeed, because even the best cultivation produces no moisture, but only conserves a part of that which falls from the clouds. Whether grapes will do depends, first, upon what the rainfall is; second, upon whether the soil is retentive; third, upon whether you cultivate in such a way as to enable the soil to exercise its maximum retentiveness. These are matters which cannot be determined theoretically - they require actual test.

Cutting Back Frosted Vine Canes.

Vines have been badly injured by the late frosts, especially the young vines which were out the most. Is there anything to be done with the injured shoots now on the vines so as to help the prospects of a crop?

If shoots are only lightly frosted they should be cut off at once as low as you can detect injury. This may save the lower parts of the shoot, from which a later growth can be made. Frosted parts ferment and carry destruction downward, and therefore should be disposed of as soon as possible. Where vines have run out considerably and badly frosted, the best practice usually is to strip off the frozen shoots so as to get rid of the dormant buds at the base, which often give sterile shoots. A new break of canes from other buds is generally more productive.

Dipping Thompson Seedless.

What is the process of dipping and bleaching Thompson seedless grapes?

One recipe for dipped raisins is as follows: One quart olive oil; 3/4-pound Greenbank soda and 3 quarts water are made into an emulsion, and then reduced with 10 gallons water in the dipping tank, adding more soda to get lye-strength enough to cut the skins, and more soda has to be added from time to time to keep up the strength. The grapes are dipped in this solution and sulphured to the proper color. This is the general outline of the process. The ability to use it well can only be attained by experience and close observation.

The Zante Currant.

Is the currant that grows in the United States in any way related to the currant that grows in Greece? If so, could it be cured like the currant that comes from Greece?

The dried currants of commerce are made in Greece and in California (to a slight extent) from the grape known as the grape of Corinth. They are not made from the bush currant which is generally grown in the United States, and the two plants are not in any way related.

Grape Vines for an Arbor.

How shall I prune grape vines, viz: Tokay, Black Cornichon, Muscat,
Thompson Seedless, Rose of Peru, planted for a grape arbor?

You can grow all the vines you mention with high stumps reaching part way or to the top of the arbor as you desire side or top shade or both. You can also grow them with permanent side branches on the side slats of the arbor if you desire. Each winter pruning would consist in cutting back all the previous summer's growth to a few buds from which new canes will grow for shade or fruiting, or you can work on the renewal system, keeping some of these canes long for quick foliage and more fruit perhaps and cutting some of them short to grow new wood for the following year's service, as they often do in growing Eastern grapes.

Pruning Old Vines.

I have some Muscat grape vines 30 years old. Can I chop off most of the old wood with a hatchet and thereby bring them back to proper bearing?

Not with a hatchet. If the vines are worth keeping at all, they are worth careful cutting with a saw and a painting of all cuts in large old wood. If the vines have been neglected, you can saw away surplus prongs or spurs, reserving four or five of the best placed and most vigorous, and cut back the canes of last summer's growth to one, two or three buds, according to the strength of the canes - the thicker the canes, the more buds to be kept. It is not desirable to cut away an old vine to get a new start from the ground, unless you wish to graft. Shape the top of the vine as well as you can by saving the best of the old growth.

Topping Grape Vines.

Is topping grape vines desirable?

Topping of vines is in all cases more or less weakening. The more foliage that is removed, the more weakening it is. Vines, therefore, which are making a weak growth from any cause whatever can only be injured by topping. If the vines are exceptionally vigorous, the weakening due to topping may be an advantage by making them more fruitful. The topping, however, must be done with discretion. Early topping in May is much more effective and less weakening than later topping in June. Very early topping before blossoming helps the setting of the blossoms. Topping in general increases the size of the berries.

Bleeding Vines.

Will pruning grape vines when they bleed injure them?

It has been demonstrated not to be of any measurable injury.

Vines and Scant Moisture.

Would it be well to sucker vines and take also some bearing canes off, or in a dry year will they mature properly as in other years if the ground is in good condition?

Vines usually bear drouth-stress better than bearing fruit trees. On soils of good depth and retentiveness, they are likely to give good crops in a dry year with thorough cultivation; still, lightening the burden of the vines is rational. Suckering and cutting away second-crop efforts should be done. Whether you need to reduce the first crop can be told better by the looks of the vines later in the season.

Sulphuring for Mildew.

For two years I have not sulphured my vineyard and had no mildew. My vines seem as healthy and thrifty as any of the neighbors' that were duly sulphured. Have I lost anything by not sulphuring?

Certainly not. In sections where mildew is practically sure to come, sulphur should be used regularly as a preventive without waiting for the appearance of the disease. There are, however, many locations, especially in the interior valley, where the occurrence of mildew is rare in sufficient volume to do appreciable harm, and then sulphuring should depend upon the weather, which favors mildew or otherwise. But be always on the watch and have everything ready to sulphur immediately; also learn to recognize the conditions under which appearances of mildew become a menace.

Grape Sugar in Canned Grapes.

How can I prevent the formation of grape sugar in canned grapes?

Take care that the syrup is of the same density as the juice of the grape when the fruit and the juice are placed together in the can. The density of the syrup and the juice are, of course, to be obtained by the use of the spindle, the same arrangement employed for determining when the percentage of sugar in the grape juice is right for raisin-making or for wine-making. Whatever the density of the juice, make the syrup the same by the use of the right amount of sugar.

Part II. Vegetable Growing

California Grown Seed.

Which are the best garden seeds to use, those raised in Ohio and the East or those raised in Washington and Oregon or those raised in this
State?

It has been definitely shown by experience and experiment that is does not matter much where the seed comes from, providing it is well grown and good of its kind. There is no such advantage in changing seed from one locality to another as is commonly supposed. Besides, it is now very difficult to tell positively where seed is grown, because California wholesale seeds are retailed in all the States you mention, and the contents of many small packets of seeds distributed in California went first of all from California to the Eastern retailers, who advertise and sell them everywhere.

Cloth for Hotbeds.

Would cloth do to cover a hotbox to raise lettuce, radishes, etc., for winter use where we get a very heavy rainfall?

Yes, if you make the cloth waterproof for its own preservation from mildew and other agencies of decay. The following recipe for waterproofing cloth is taken from our book on "California Vegetables": Soften 4 1/2 ounces of glue in 8 3/4 pints of water, cold at first; then dissolve in, say, a washboiler full (6 gallons) of warm water, with 2 1/2 ounces of hard soap; put in the cloth and boil for an hour, wring and dry; then prepare a bath of a pound of alum and a pound of salt, soak the prepared cloth in it for a couple of hours, rinse with clear water and dry. One gallon of the glue solution will soak about ten yards of cloth. This cloth has been used in southern California for several years without mildewing, and it will hold water by the pailful. Where the rain is heavy and frequent, the cloth should be well supported by slats and given slope to shed water quickly. Of course, this is only a makeshift. Glass would be more satisfactory and durable in a region of much cloudiness and scant sunshine; the greater illumination through glass will make for the greater health and growth of the plants.

Soil for Vegetables.

Some of my soil bakes and hardens quickly after irrigation, but I have an acre or so of sandy soil. Would this be best for garden truck and berries?

Sandy, loamy soil is better than the heavy soil for vegetables and berries, if moisture is kept right, because it can be more easily cultivated and takes water without losing the friable condition which is so desirable. A heavier soil can, however, be improved by the free use of stable manure or by the addition of sand, or by the use of one or more applications of lime at the rate of 500 pounds to the acre, as may be required - all these operations making the soil more loamy and more easily handled.

Vegetables in a Cold, Dark Draft.

What vegetables will thrive in localities where the sun shines only part of the day? I have a space in my garden that gets the sun only between the hours of 11 and 5, thereabouts; I would like to utilise those places for vegetables if any particular kind will grow under such conditions. The soil apparently is good, of a sandy nature, with some loam. The place is high and subject to much wind.

You can only definitely determine by actual trial what vegetables will be satisfactory under the shade conditions which you describe. You may get good results from lettuces, radishes, beets, peas, top onions, and many other things which do well at rather a low temperature, while tomatoes, peppers, eggplant, etc., would probably be worthless. Your soil is probably satisfactory and you can easily keep the moisture right by being careful not to use as much water as you would in open sunshine. The behavior of the plants will be directly dependent upon the temperature and the sunshine which they receive under the conditions described.

Jesusalem Artichokes.

What is the best time for planting Jerusalem artichokes?

Jerusalem artichoke tubers are planted in the spring after the ground has become warm and the heavy frosts are over. The planting may be done in rows far enough apart for cultivation, the tubers being set about a foot apart in the row. This tuber grows like a potato, but is more delicate than the potato. It is inclined to decay when out of the ground, but will not start growth as early as the potato, and therefore it is not desirable to start it early in the winter if the winters are cold and the ground apt to be very wet. Do not cut the tubers for seed as you would potatoes.

Globe Artichokes.

I have land that will grow magnificent artichokes. Two plants last year (variety unknown) produced heavy crops of buds, but the scales opened too wide and allowed the center to become fibrous and were unsalable. Is this due to climate, lack of sufficient water, or to not having the right variety?

Many artichokes which are planted should really be put in the ornamental class - they are either a reversion from a wilder type in plants grown from the seed or they never have been good. In order to determine which varieties you had better grow on a large scale, it is desirable to get a few plants of the different varieties as offered by seedmen. In this way you would find out just what are considered best in different parts of the State, and propagate largely the ones which are best worth to you. By subdivision of the roots you get exactly the same type in any quantity you desire - ruling out undesirable variations likely to appear in seedlings.

Artichoke Growing.

Is the Globe artichoke a profitable crop to raise commercially? Near Pescadero a company has been formed to raise it for Eastern shipment. Is it a very profitable crop to raise? Are certain varieties worthless?

Considerable quantities of Globe artichokes are grown in southern and central California for Eastern shipment. There is a limit to the amount which can be profitably shipped, because people generally, at the East, do not know the Globe artichoke and how to eat it, but more of them are learning the desirability of it every year. There are species which are only ornamental, as a bad weed.

Asparagus Growing.

What is the average commercial yield of asparagus to the acre in California? Also, how long it takes asparagus to come into full bearing, and what yield could be expected after two years' growth? Is asparagus resistant to moderate quantities of alkali in the soil?

The yield of asparagus is from one to four tons of marketable shoots per acre, according to age and thrift of plants, etc., the largest yields being on the peat lands of the river islands. On suitable lands one ought to get at least two tons per acre. Roots may yield a few days' cuttings during their second year in permanent place; the third year they will stand much more cutting, and for several years after that will be in full yielding. Asparagus enjoys a little salt in the land, but one would not select what is ordinarily called "alkali land" for growing it - not only because of the alkali but because of the soil character which it induces.

Bean Growing.

We have a small field of beans, and would like to know which is the best and most profitable way to crop them.

Cultivate the beans so that the plants may have plenty of moisture to fill the pods, then let them dry and die. Gather the dry plants before the pods open much, and let them dry on a clean, smooth piece of ground or on the barn floor. When they are well dried, thresh with a flail, rake off the straw, sweep up the beans and clean by winnowing in the wind or with a fanning mill with suitable screens.

Hoeing Beans.

Should beans be hoed while the dew is on the vine?

Beans had better be hoed with the dew on them than not hoed at all. The only objection to hoeing with the dew on is that the hoer will get his feet wet, the vines will become untidy from adhering dust, with a possible chance of the leaves becoming less effective and the pollination of the blossom rendered less liable to occur.

Beans as Nitrogen Gatherers.

I grow string beans in my rotation to restore nitrogen, but I see it stated that not all beans are valuable for this purpose. Are the common bush varieties nitrogen gatherers?

Probably they are all doing it in various degrees. Pull up or dig up a few plants when growing actively, not too early nor too late in the season, and look for nodules on the roots. Number and size considered together will measure their activity in this line in your soil.

Bean Growing.

I want to plant beans of different varieties. The land is rich, black loam with a little sand. When is the best time to plant? If planted early, what shall we do to keep the weevils out of them?

It is desirable to plant beans as early as you can without encountering danger of frost killing. No particular date can be mentioned for planting because the dates will vary in different locations according to the beginning of the frost-free period. The best way to escape weevil is to sell most of the beans as soon as harvested, treating those which you retain for seed, or for your own use, with bisulphide of carbon vapor or by gently heating to a temperature not above 130 degrees, which, of course, must be done carefully with an accurate thermometer so as not to injure germinating power. Unless you know that beans do well in your locality, it would be wise to plant a small area at first, because beans are somewhat particular in their choice of location in California, and one should have practical demonstration of bearing before risking much upon the crop.

The Yard-Long Bean.

I wish to ask about the very long bean which I think was introduced from China into California. I remember seeing one vine when I was living in California which I think must have been 20 or 30 feet long and had hundreds of pods and each of these pods were from 2 to 3 feet long. Are these beans generally considered eatable? Would they be at all suitable to get as a field bean which the hogs eat?

You probably refer to the "yard-long" pole bean. It is a world variety and may have come to California from China as you suggest, but it has also been well known for generations in Europe and was brought thence to the Eastern States at some early date. It is generally accounted as an unimportant species and certainly has not risen to commercial account in California. The beans are edible and the whole plant available for stock feeding, but there is no doubt but that the growth of some of the cowpeas would be preferable as a summer field crop for hog pasture.

Why the Beans are Waiting.

Can you tell me why pink beans which were planted early in Merced county, irrigated four times, hoed four times and cultivated, have no beans on them? The vines look finely.

Probably because you had too much hot, dry wind at the blooming. This is one of the most frequent troubles with beans in the hot valley, but the pink bean resists it better than other varieties. As the heat moderates you are likely to get blossoms which will come through and form pods, and then the crop will depend upon how long frost is postponed. You have also treated the plants a little too well with water and cultivation. You had better let them feel the pinch of poverty a little now; they will be more likely to go to work.

Blackeye Beans.

What is the best way to prepare land for Black-eye beans? How much seed is required per acre, and what is the estimated cost of growing them? The soil is a well-drained clay loam.

The cost of growing is not particularly different from other beans, and will vary, of course, according to the capacity and efficiency of the plows, harrows, teams, tractors, men, etc. Every man has to figure that according to his conditions and methods of turning and fining the land. Sow 40 pounds per acre in drills 3 feet apart, and cultivate as long as you can without injuring the vines too much. Sowing must of course be done late, after the ground is warm and danger of frost is past, though the plowing and harrowing should be done earlier than that.

Blackeye Beans are Cow Peas.

I sent for some Blackeye cow peas; they look like Blackeye beans. Am sending you a sample of what I got. What are they?

Yes, they are in the cow pea group, but there are other cow peas which would not be recognized as having any relation to them. All cow peas are, however, beans, and they have not much use for frost. They are not hardy like the true pea group.

Growing Horse Beans.

Does the soil need to be inoculated for horse beans? I intend to plant five acres about January 1, on the valley border in Placer county and they get heavy frost in the morning. Does frost hurt them? How shall I plant them?

California experience is that horse beans grow readily without inoculation of the seed. Quite a good growth of the plant is being secured in many parts of the State, particularly in the coast region where the plant seems to thrive best. It is one of the hardiest of the bean family and will endure light frost. How hardy it will prove in your place could be told only by a local experiment. Whether it can be planted after frost danger is over, as corn is, and make satisfactory growth and product in the dry heat of the interior summer must also be determined by experience.

The horse bean is a tall growing, upright plant which is successfully grown in rows far enough apart for cultivation, say about 2 1/2 feet, the seed dropped thinly so that the plants will stand from 6 inches to 1 foot apart in the row.

Growing Castor Beans.

Give information on the castor oil bean; the kind of bean best to plant, when to plant and harvest, the best soil, and where one can market them.

Castor bean growing has been undertaken from time to time since 1860 in various parts of California. There is no difficulty about getting a satisfactory growth of the plant in parts of the State where moisture enough can be depended upon. Although the growing of beans is easy enough, the harvesting is a difficult proposition, because in California the clusters ripen from time to time, have to be gathered by hand, to be put in the sun to dry, and finally threshed when they are popping properly. The low price, in connection with the amount of hand work which has to be done upon the crop, has removed all the attractions for California growers. There is also, some years, an excess of production in the central West, which causes prices to fall and makes it still more impracticable to make money from the crop with the ordinary rates of labor. The oil cannot be economically extracted except by the aid of the most effective machinery and a well equipped establishment. Oil-making in the rude way in which it is conducted in India would certainly not be profitable here.

Legume Seed Inoculation.

Is there any virtue in inoculating plants with the bacteria that some seed firms offer? I refer to such plants as peas and beans.

If the land is yielding good crops of these plants and the roots are noduled, it does not need addition of germs. If the growth is scant even when there is enough moisture present and the roots are free from nodules, the presumption is that germs should be added. Speaking generally, added germs are not needed in California because our great legume crops are made without inoculation. Presumably, burr clover and our host of native legumes have already charged the soil with them. If, however, such plants do not do well, try inoculation by all means, to see if absence of germs is the reason for such failure or whether you must look for some other reason. If the results are satisfactory, you may have made a great gain by introduction of desirable soil organisms which you can extend as you like by the distribution of the germ-laden soil from the areas which have been given that character by inoculation of the seed.

Beans on Irrigated Mesas.

Would white and pink beans do well on the red orange land at Palermo with plenty of water? I have in mind hill land, the hills being very red and running into a dark soil in the lower part. How many beans could I get per acre?

Probably nothing would be better for the land or for the future needs of the trees than to grow beans. An average crop of beans, for the whole State and all kinds of beans, is about one ton to the acre. What you will get by irrigation on hot uplands we do not know. Beans do not like dry heat, even if the soil moisture is adequate. They do not fructify well even when they grow well. The pink bean does best under such conditions. All beans, except horse beans, must be brought up after frost dangers are all over, and this brings them into high heat almost from the start in such a place as you mention. You should find out locally how beans perform under such conditions as you have, before undertaking much investment.

Leases for Sugar Beets.

I have land in Yolo county that has made an average yield yearly of from 12 to 18 sacks of wheat and barley. A beet sugar company proposes renting this land and plant it to sugar beets and I would prefer not to consider any agreement of less than five years' duration. The particular point that I would like to have you advise me on is the effect sugar beet has upon the soil.

You certainly have good soil, and it is not strange that a sugar company should desire to rent it for its purposes. There is, however, a great question as to whether it would be desirable to run to beets continually for five years. Beets make a strong draft on some components of the soil, and it is a common experience that they should not be grown year after year for a long period, but should take their place in a rotation, in the course of which one or two crops of beets should be followed by a crop of grain, and that if possible by a leguminous plant like alfalfa or an annual legume like burr clover used for pasturage, and then to beets again. Beets improve soil for grain, because of the deep running of the root, and because beet culture is not profitable without deep plowing and continuous summer cultivation. This deepens and cleans the land to the manifest advantage of the grain crop, but still the beet reduces the plant food in the soil and some change of crop should be made with reference to its restoration. We would much prefer to lease it for two years than for five years of beet growing.

Topping Mangel Wurzels.

Does it harm the mangel wurzels if their tops ore cut off once a month?

Removing leaves will decrease the size and harden the tissues of the beet root. If you wish to grow the plant for the top, the root will continue to put out leaves for you for a time; if you grow it for the size and quality of the root, you need all the leaf-action you can get, therefore do not reduce the foliage.

Blooming Brussels Sprouts.

Are Brussels sprouts male and female? Some of my plants are flowering and show no signs of sprouts, while those that are not, show some small eyes at stem that look like young sprouts.

Brussels sprouts ought to form the sprouts without flowering, just as a cabbage heads without flowering. Those plants which show flowers have been stopped by drought or otherwise, and have taken on prematurely the second stage of growth which is productive of seed and is undesirable from the point of view of growing heads.

Blanching Celery.

I desire to know the different methods by which the celery is bleached, and particularly whether boards or other material other than earth is used for this purpose.

There is some blanching of celery with boards, cloth wrappings, boot-legs, old tiles, sewer pipes, etc., in market gardens in different parts of the State, but the great commercial product of celery for export is blanched wholly by piling the light, dry earth against the growing plant. As we do not have rains during the growing season and as the soil on which celery is chiefly grown is particularly coarse in its texture, there is no rusting or staining from this method of blanching. It shakes out clean and bright. Conditions which make earth-blanching undesirable in the humid region do not exist here.

Corn in the Sacramento Valley.

Is it practical to raise corn in the Sacramento volley? Are the soil and climatic conditions suitable?

The success of corn on plains and uplands in the Sacramento valley has not yet been fully demonstrated, although good corn is grown on river bottom lands, and it is possible that much more may be done with this grain in the future than in the past. Corn does not enjoy the dry heat of the plains, and even when irrigated seems to be dissatisfied with it. How far we shall succeed in getting varieties which will endure dry heat and still be large and productive will ere long be determined by the experiments which are in progress. The old Sacramento valley farmer has been justified to some degree in his conclusion that his is not a corn country. Still it may appear so later.

Plant Corn in Warm Ground.

I also put in a lot of corn and none of it came up. The ground was damp and rather cold, as well as being alkali.

Corn should never be planted in cold, wet ground - in fact, very few seeds should be. Besides, corn has no use for alkali.

Sweet Corn in California.

I have been informed that sweet corn cannot be raised in this part of the country, an account of worms eating the kernels before the ear has matured. Is there any method of overcoming this difficulty?

You have been correctly informed concerning the difficulty in growing sweet corn. Although many experiments have been made, no method of overcoming this pest has yet been demonstrated. For this reason canning of corn is not undertaken in this State, and for the same reason most of the green corn ears sold in our markets have the tops of the ears amputated. It is sometimes possible to escape the worm by planting rather late, so that the ears shall develop after the moth, which is parent of the worm, has deposited its eggs.

Forcing Cucumbers.

Give information on growing hot-house cucumbers, and also if you think it would pay me to go into the business in southern California.

Forcing of cucumbers has been undertaken for a number of years in California and formerly was considered unprofitable because cucumbers grown in the open air in frostless places came in before the forced product could be sold out at sufficiently high prices to make the venture profitable. Recently, however, owing to our increased population in cities and larger demand of products out of season, forcing becomes more promising and is worthy of attention. Forcing of cucumbers in California can be done at very much less expense, of course, than elsewhere, because of the abundance of winter sunshine and the fact that sufficiently high temperatures can be secured in glass houses with exceedingly little if any artificial heat: The chances of growing cucumbers out of season for shipment eastward and northward can be discussed with the officers of the California Vegetable Growers' Union, which has offices and warehouse in Los Angeles.

Cucumber Growing.

I have a piece of red so-called orange land which has produced excellent wheat. Will you give information about its adaptability to cucumbers? Are there pickle factories in the State which would demand them in quantities, and is there much other demand for them? About when should they be planted, and how much water would they need?

The cucumber needs a retentive soil which does not crack and bake, and such a soil is made by abundance of organic matter. Your orange soil, unless heavily treated with stable manure and given plenty of time for disintegration, would probably give you distressful cucumber plants, if it has come right out of wheat-growing. Besides, cucumbers do not like dry heat, even if the soil be kept moist by irrigation. Oranges will do well under conditions not favorable to cucumbers. Cucumber plants must come up after danger of frost is over. The amount of water they require depends upon how moist the soil is naturally, and as the crop is chiefly grown on moist river lands and around the bay, it is chiefly made without irrigation. Such lands have a cucumber capacity equal to the consumption of the United States, probably, and the pickle factories can usually get all they can use at a minimum transportation cost. Large-scale plantings should only be made by men who know the crop and have definite information or contract for what they can get for it.

Ginger in California.

We have ginger roots in a growing condition with sprouts and bulbs growing an them, but we do not understand how to raise the plants.

Growing ginger in California in a commercial way has not been worked out, although roots have been introduced from time to time. Plant your roots in the garden, just as you would callas, where you can give them good cultivation and water, as seems to be necessary, and note their behavior under these favorable conditions before you undertake any large investment in a crop.

Licorice Growing in California.

I have for some time been seeking far some information as to the method of preparation for market and sale of licorice roots. I have a lot of them and have never been able to find a market, and do not know how they are prepared for market.

Licorice was first planted in California about 1880 by the late Isaac Lea, of Florin, Sacramento county. Mr. Lea grew a considerable amount of licorice roots and gave much effort to finding a market for it. He found that the local consumption of licorice root was too small to warrant growing it as a crop; that the high price of labor in digging the roots, and the high cost of transportation of the roots to Eastern markets would make it impossible for him to undertake competition in the Eastern markets with the Sicilian producers, unless, perhaps, he could build an extracting factory and market licorice extract, the black solid which is sold by the druggist, and which the Sicilians produce in large quantities. The preparation of licorice root is simply digging and drying, but the preparation of the extract requires steam extractors and condensers. California could produce licorice, for we have a good climate for it. If it is grown on light, sandy loams, it could be pulled from the ground by the yard at rather small expense, and yet, one should not undertake the production unless he wished to put in much time and money in working up economical production and marketing in competition with the foreign product, produced by cheap labor and with the advantage of processes well known and established by long usage. Experiments should be circumspectly undertaken, for licorice is one of the worst weeds in the world, and extremely difficult of eradication probably.

Growing Lentils.

Give information regarding the planting and raising of lentils. Can they be grown in the Sacramento valley in the vicinity of Colusa, and at a profit?

Lentils are as easily grown in California as common peas, and will do well as a field crop if started during the rainy season, as they are hardy enough to survive our ordinary valley frosts. With respect to lentils, it may be said that excellent as these legumes are for many purposes, they do not seem to be well known to American consumers, and therefore the amount to be grown is limited, until you know who will buy larger quantities of them at a good price.

Canada Peas for Seed.

I want to raise Canada peas for the seed. In what month of the year is the best time to plant them; also how many pounds to the acre to be sowed broadcast on rolling land in Napa?

Broadcast from 80 to 100 pounds of seed per acre as soon as you can get the ground into good condition. What you get will depend much upon how late spring rains hold this year. We should only try a small area this year to see what happens, for you probably should have started earlier in the season. On uplands it will always be a question whether your soil will hold moisture enough to mature a good seed crop.

Growing Niles Peas.

How shall I plant and handle a crop of Niles peas?

Niles peas are hardy and will make a good crop on any good soil, if planted early in the season so as to make the main part of their growth before the heat of the summer comes on. Under garden conditions they can, of course, be grown all summer.

Transplanting Lettuce.

I have lettuce plants that have been transplanted to head. Occasionally I find a head that has withered away and upon examining it find it rotted away at the stem. Can you suggest a remedy for it?

Your lettuce plants are destroyed by the "damping, off" fungus. It would be preventable by reducing the amount of moisture until the transplanted plant had opportunity to re-establish itself in the soil and thus come into condition to take water. The chance of it could also be reduced by using a certain amount of sand in connection with the soil, unless it is already very sandy, and by a shallow covering of sand on the surface around the plants after they are reset, in order to prevent too great accumulation of moisture.

Handling Winter Melons.

Give particulars regarding harvesting, storaging, and shipment of winter melons. How do you harvest and pack them for distant market?

There is no particular system in the handling of winter melons. They are gathered into piles on ground where water will not gather and covered with the trash of the vines on which they grow. They will keep for months in this way, as our autumn temperatures do not freeze them. Other growers collect them in open sheds shaded from sun and rain, and still others put them into barns or shallow cellars under buildings, etc. The melons are very durable and seem disposed to keep in any old way. The melons are shipped in large packing cases with slat sides, or in the smaller slat crates that are used for summer cantaloupes. No packing is used, generally. If it seemed necessary, a little clean straw would be sufficient.

Ripe Melons.

How can I tell when a watermelon is fully ripe? What is the method used by growers in picking for commercial shipping?

Gently press the sides of a melon and if it crackles a little bit, all right; if it makes no sound then go to another. Commercial pickers look at the little spiral between the melon and the nearest leaf. If it is withered they pick the melon, if fresh, pass it until next picking.

Growing Onion Seed and Sets.

Will you give localities of the leading production of onion seed or dry sets in your State?

Onion seed is grown in several parts of the State, largely in the Santa Clara valley adjacent to the city of San Jose. Onion sets are largely produced in Orange county, near Los Angeles, for eastern shipment, for which purpose they are grown under contract.

Ripening Onions.

I am raising some onions from bottom sets and as they are growing nicely and are beginning to swell at the bulb some advise me to cut the tops off and some advise me to bend them over or tramp them down.

Do not cut off the tops of the onions. If they seem to be overgrowing and not disposed to ripen the bulb, the top can be broken down, thus partly arresting the vegetative energy of the plant and causing maturity.

Onions from Sets.

Will onion sets planted in July grow and mature in the fall months?

Good onion sets grown during the winter and spring should be mature by July and if planted after drying would proceed to make a full growth of large onions if growing conditions should be right for them, that is, the soil moist and the temperature not too high.

How Many Crops of Onion Seed?

Does the growing of onion seed exhaust adobe land, and if so, how many years' cropping before it requires rest or fertilizing?

The growth of any seed crop, including cereal grains, of course, makes a supreme draft upon soil fertility. How long a certain soil can stand it, depends upon the amount of fertility it has when the draft begins. The best rough way to tell how it is going, is to watch the growth and crop, when moisture conditions are known to be favorable. If you get a good growth of the plant it is still good to make the seed.

Onions from Seed.

Will onions from seed mature the same season if they are irrigated? Some tell us they will not, so we would be very much pleased to hear from you.

Onions grown from the seed do fully develop during the growing season following the planting of the seed. In fact, nearly all California onions are grown in that way. Our growing season is so long that we do not need to use onion sets to any extent, as they do in short-summer climates.

Dry Farming with Chili Peppers.

If I set chili pepper plants down six or eight inches lower than the surface of the ground and fill in as the plants grow larger, will this help in case I could not get water enough? My soil is a deep sandy loam. We have had between five and six inches of rain. Do you think water every fifteen days would be enough?

On such light soil as you mention, the plants can be planted deeply and a certain amount of soil brought up to the plants by cultivation without injury. As this plant has a long growing season and matures its crop rather late, you will undoubtedly need irrigation. Probably irrigation twice a month will be sufficient in connection with good cultivation, but you will have to watch the plants and apply the water as it seems to be needed, rather than by a specific scheme of days.

Harvesting Peanuts.

I would like information regarding the curing of peanuts. Should they be bleached, and, if so, how is it done? Does bleaching affect the keeping qualities?

It is not usual to bleach peanuts. They should be grown in such light soil that they will not be stained, and the common method of curing is to dig or plow up, throw the vines, with nuts attached, into windrows and allow them to lie a week or ten days for drying. Then the nuts are picked into sacks and cleaned before shipment in revolving drums, followed by a grain fan which throws out the light nuts and other rubbish. Bleaching would not destroy the keeping quality probably, but it would destroy the flavor and the germinating power. The latter would not matter, except with such nuts as you wish to keep for seed, because the roasting destroys the germinating power also, but sulphuring, which would reduce the flavor, would give the product a bad name. Possibly some growers do bleaching, but, if so, they have to be pretty careful about it. The cost of the operation would also be a bar to profit, for peanuts are grown on a narrow margin owing to competition with importations grown with cheap labor.

Adobe and Peanuts.

Is adobe land good for the peanut? Is it harder to start than in other soils or not?

It is not good at all. Peanuts require the finest, mellowest loam with sand enough to prevent crust, and moisture even and continuous. The surface must be kept loose so that the plant can bury its own bloom stem and the under soil light and clean so that it will readily shake from the nuts and not stain them. Adobe is the worst soil you could find for peanuts.

Cutting Potatoes.

What would be the most profitable potato to plant in the Salinas valley, and how small can a potato be cut up for planting? How many eyes should each piece contain in order to make a good growth and be profitable?

Probably the best potato for your district would be the Burbank, which is largely grown near Salinas and brings the highest price. It is customary to cut a medium-sized potato in two pieces and a large one in four pieces. One can be very economical of seed by smaller cutting, but it would require the most favorable conditions to bring a vigorous growth. Probably pieces weighing not less than two ounces would be best under ordinary conditions. Potatoes which are rather small may be used for seed if well matured and have good eyes. It is dangerous, however, to use the small stuff - too small for sale. Unless the soil and moisture conditions are extra favorable, the growth will be weak and unsatisfactory.

Potato Planting.

How many sacks of potatoes are to be planted to an acre, and how many eyes are to be left in a seed? If, for instance, we plant seed with three eyes, how many potatoes should we get from that vine?

Potatoes are planted all the way from five to fifteen sacks to the acre, probably about ten sacks being the average. There is no particular number of eyes specified in preparing the seed, according to common practice. Good medium-sized potatoes are generally cut in two pieces crosswise, and large potatoes in four pieces, cutting both ways. There is no definite relation between the number of eyes planted and the number of potatoes coming from them. This has been the subject of innumerable experiments, and the conclusion is that the crop is more dependent upon good soil and favorable growing conditions than upon any way of preparing the seed.

Northern Potatoes for Seed.

Do you regard northern-grown seed potatoes sufficiently better to make it worth while paying freight on them from the State of Washington?

Experience seems to indicate the superiority of northern-grown seed potatoes, not only in this State, but on the Atlantic Coast, and they are largely depended upon. Systematic demonstration by comparative tests has been made by the Vermont station and preference for northern-grown seed seems, to be justified.

Potato Planting.

I have ten acres of land in Placer county which I propose to put into potatoes next spring. It has been recommended to me to put potatoes in as early as January. It seems to me that January is rather early; however, it is said that this land is in the orange belt and practically free from frost.

Whether you can plant potatoes to advantage in January or not depends upon the temperatures which you are likely to meet after that date, also whether the ground is warm enough in January, because there is no advantage in planting in cold ground nor in soil that is too wet at the time. The earliest potatoes, of course, come from planting much earlier than January; usually as soon as the ground is moistened enough in the autumn. The potato will stand some frost, but autumn planting is not feasible in places which are under hard freezing or receive too much cold rain water.

Potatoes Should be Planted Early.

I have Early Rose potatoes planted about May first. The tops look fine, but there are few potatoes and small, and, though not developed, have commenced growing a second time, sprouts starting from the new potatoes. When should I plant and what care should they have?

Your potatoes act peculiarly because of intermittent moisture - the plant being arrested by drought and then starting again, which is very undesirable. To avoid this, potatoes should be planted earlier so as to get a large part of their growth during the rainy season. If planted late the ground should be well wet down by irrigation, and then plowed and cultivated, and irrigation should be used while the plant is growing well. If this is done, potatoes can be successfully grown by irrigation, but if the land is allowed to become dry the plant is arrested in its growth for a time and a second and undesirable growth is started.

Potato Balls.

I find in potato writings of forty years ago that the seed from the potato balls which form on the tops of the plants is recommended for growing the best potatoes. In later books I find no mention of them and all are advised how to cut the tubers to get seed potatoes.

The seed of the potato plant which is found in the "balls" which develop on the tops of the plant is only valuable for the origination of new varieties, with the chance, of course, that most of them will be inferior to the tubers produced by the plant which bears the seed. Therefore, these seeds are of no commercial importance. There has also sometimes developed upon the top of the plant what is called an aerial tuber, which is even of less value than the seed ball, because it does not contain seed nor is it good as a tuber.

Forty years ago there was a great demand for newer and better kinds of potatoes which has, since that time, been largely supplied, and commercial potato-growing consists in multiplying the standard varieties which best suit the soil and the market. This is done by planting the tuber itself, which is really a root-cutting and therefore reproduces its own kind. Those who are originating new kinds of potatoes still use seed from the balls, either taking their chances by natural variation or, by hybridizing the blossoms, increasing the chances for variation from which desirable varieties are taken by selection, to be afterward multiplied by growth from the tubers.

Seed-Ends of Potatoes.

Is it bad practice to plant the seed-ends of potatoes?

The seed-end of the potato is the least valuable part of it, but it is better probably to plant than to reject it.

The Moon and Potato Planting.

Is there any foundation to the oft-repeated story about potatoes in the light of the moon running to tops and the dark of the moon to spuds?

If we paid any attention to the moon in planting, we should plant in the dark of the moon so as to give the plant opportunity to make use of whatever additional light the full moon afforded.

Planting Whole Potatoes.

One man states the only way to cut seed is to take a potato and cut the ends off and not divide the potato any more; or, in other words, a whole potato for each seed.

Good results are obtained by planting whole potatoes, but in that case there is no advantage in removing the ends.

How to Cut Seed Potatoes.

Would it pay in returns to use large potatoes for seed in preference to culls?

Large potatoes are better than culls, but medium-sized potatoes are better than either. Many experiments have been made to determine this. At the Arkansas station whole tubers two to three inches in diameter yielded 18 per cent more than small whole tubers three-quarters to one and one-quarter inches in diameter, and large cut tubers yielded 15.8 per cent more than small cut tubers.

Cutting Potatoes to Single Eyes.

Some say only one eye to a piece; others say several eyes - which is better?

In one experiment potatoes cut to single eyes with each piece weighing one-sixteenth of an ounce yielded 44 bushels to the acre, while single eyes on two-ounce pieces yielded 177 bushels to the acre. Experiments in Indiana showed that the yield usually increased with the weight of the set and that the exact number of eyes per cutting is relatively unimportant.

Potato Scab.

Can potatoes be treated in any way before planting to prevent the new ones from being what is called "scabby"?

There are two successful treatments for scab in potatoes. One is dipping in a solution of corrosive sublimate. Dissolve one ounce in eight gallons of water and soak the seed potatoes in this solution for one and one-half hours before cutting. This treatment kills the scab spores which may be upon the exterior of the potatoes. More recently, however, to avoid danger in handling such a rank poison as corrosive sublimate, formaldehyde has been used, and one pint of commercial formaldehyde, as it is bought in the stores, is diluted with thirty gallons of water, and potatoes are soaked in this for two hours. Thirty gallons of this dip ought to treat about fifty bushels of potatoes.

Double-Cropping with Potatoes.

I am told that here two crops of potatoes can be raised by planting the second crop in August. I have five acres which will be ready to dig in July. Can I dig these Potatoes and use them for seed at once for another crop, or won't they grow? I have a crop of barley, and as it is heading out now, I want to put potatoes on the ground after I take the barley off. I have plenty of water to irrigate.

If your potatoes ripen in July and you allow those which you desire for seed to lie upon the ground and become somewhat greenish, they are likely to sprout well for a second crop. They should not, however, be planted immediately. Whether you get a second crop successfully or not depends upon how early the frosts come in your district. Whether you get potatoes after barley or not depends also upon how much moisture there remains in the soil. By irrigating thoroughly after harvesting the grain and then plowing deeply for potatoes, you would do vastly better than to plant in dry ground and irrigate afterward.

When to Plant Potatoes.

I have been puzzled to understand Potato growing in California. Do you have more than one cropping season, and if so, about what dates are they due?

Every month in the year potatoes are being put into the ground and being taken out of the ground somewhere in California. We have, then, practically a continuous planting and harvesting season. There is, however, a division possible to make in this way: Plantings undertaken in September and October are for winter supplies of new potatoes, which begin about the holidays and continue during the winter. There is also in southern California a planting beginning in January, which might be called the earliest planting for the main crop, and other plantings for the main crop in the central and northern parts of the State begin in February and continue until May, according to the character of the land; that is, whether it is upland, on which the planting is earlier, or whether it is lowland along the rivers where excessive moisture may render the land unsuitable until April or May. The harvesting of the main crop then begins in May and continues during the whole of the summer, according to the character of the land cropped over, lapping the planting time for early potatoes first mentioned. It is also true by use of properly matured seed one can secure, in some places, two crops a year, if there is sufficient inducement therefor. Thus it comes about that we are continually planting and digging potatoes according to local conditions and the possibility of selling advantages.

Keeping Potatoes.

Advise me how to keep my potatoes. What is the best way? Would a dark room be suitable? Some people are digging holes in the ground to put them in.

Potatoes, if properly matured and free from disease, will keep for a considerable time in dark rooms kept as cool as possible. They must be kept away from the reach of the moth, which is parent to the worm producing long black strings inside of the potato. If they are thoroughly covered with boards or sacking or straw, so as to keep the moth from reaching the potato, they may be held for a long time in the open air, and covering with earth, as your neighbors are doing, will be all right until the rains come and cause decay by making the soil too wet. The main point is to keep the tubers as cool as possible and out of reach of the potato moth.

Potato Yield.

What is the yield per acre of potatoes on the best land around Stockton, Cal., where work is done properly; also what is the yield for potatoes along the coast?

The average yield of potatoes in California, taking the whole acreage and product as reported by the last United States census, is 147 bushels to the acre. In Stockton district, on good new reclaimed land the yield has been reported all the way from 300 to 800 bushels per acre - the crop declining rapidly when continued on the same land. One year's crop in the Stockton district was estimated at 45,000 acres averaging 125 sacks per acre. The coast yield would be more like the general average for the State as first given.

New Potatoes for Seed.

Can I plant American Wonder potatoes for the first crop, and let enough of them mature to use for seed for the second crop, to be planted the first or middle of July?

It is possible to use potatoes grown the same year as seed for the later crop, providing you let the potatoes mature first by the complete dying down of the vines, and second by digging the potatoes allow them to lie in the open air, with some protection against sunburning, until the potatoes become somewhat greenish. If this is the case the eyes will develop and seed will grow, while without such treatment you might be disappointed in their behavior. Of course, the question still remains whether it would be desirable to do this or to plant some later variety earlier in the season when the growing conditions would be better.

Potato Growing.

In what locality are the best early potatoes grown in California? Can they be raised on wheat lands without irrigation as an early crop?

Early potatoes are grown in regions of light frosts in all parts of the State - around the bay of San Francisco, on the mesas in southern California, and to some extent at slight elevations in the central part of the State. The potato endures some frost, but one has, for an early crop, to guard against the locations subject to hard freezing. Most of our potatoes are grown without irrigation because, on uplands, winter temperatures favor their growing during the rainy season. The middle-season and late potatoes are grown on moist lowlands where irrigation is not necessary. In proper situations, much of the land which is used for potatoes has at some time produced wheat or barley, corn or sorghum, and other field crops.

Potatoes After Alfalfa.

I have been a successful potato grower in Ohio. I have the best alfalfa soil and it is now in its fourth year of productiveness in that
crop. I would like to grow potatoes in a small way.

Proceed just as you would at the East in getting potatoes upon a
red clover sod. Turn under the alfalfa deeply now if the soil will
work well, and roll your sandy soil. You must use a sharp plow to
cut and cover well. If there is moisture enough the alfalfa, plowed
under in the fall, ought to be decayed by February, when you could
plant potatoes safely, probably, unless your situation is very frosty.
If you plant early you ought to get the crop through without irrigation if you cultivate well and keep the land flat.

Flat or Hill Culture for Potatoes.

Is it better to hill potatoes or not?

During the dry time of the year potatoes should be grown with
flat cultivation, except as it may be necessary to furrow out between
the rows for the application of irrigation water. Potatoes grown
during the rainy season in places where there is liable to be too
much water, can often be hilled to advantage, but dry-season cultivation of practically everything should be as flat as possible to retain moisture near the surface for the development of shallow-
rooting plants.

Bad Conditions for Potatoes.

Our potatoes were planted early and were frosted several times
while young. As we come to harvest them we find them with very
large green tops but the potatoes are about the size of a hen's egg
and from that they run down to the size of a pea. The larger ones

are beginning to send out roots, four or five to a potato. The potatoes have not been irrigated lately and the ground they are in is dry.

The ugly behavior of your potatoes is doubtless due to irregularities in temperature and moisture which have forced the plants into abnormal or undesirable activity. Potatoes should have regular conditions of moisture so that they shall proceed from start to finish and not stop and start again, for this will usually make the crop unsatisfactory and worthless. Excessive moisture is not desirable, but the requisite amount in continuous supply is indispensable.

Potatoes on Heavy Land.

Will potatoes grow well in adobe land, or partly adobe, that has not been used for seven years except for pasturing?

Although potatoes enjoy best of all a light loam in which they can readily expand, it is possible to get very good results on heavy land which has been used for pasturage for some years, providing the land is broken up early and deeply and harrowed well in advance of planting and thorough cultivation maintained while the crop is growing. The content of grass roots and manure which the land has received during its period of grazing tends to make the soil lighter and will also feed the plant well. For this reason better potatoes are had on heavy land after pasturage than could be had on the same land if continually used for grain or for some other crop which tended to reduce the amount of humus and to make the land more rebellious in cultivation.

Storage of Seed Potatoes.

We need potatoes for late planting and have found a good lot which is being held in cold storage at temperatures from 34 to 36 degrees F. They have not been there long, however. Would that hurt them for seed, and also how long could they be safely left there now before planting?

Seed potatoes would not be injured in storage, providing the temperature is not allowed to go below the freezing point. They should not, however, be allowed to remain longer in storage, but should be exposed to the sun for the development of the eyes, even to the sprouting point being desirable before planting. The greening of the potato by the sun is no disadvantage. We would not think of planting potatoes directly from storage, because, owing to the lack of development in the eyes, decay might get the start of germination.

Potatoes and Frosts.

Can I keep frost off of potato tops by building smudge fires! I would like to plant about February 1, but we usually have a few light frosts here during March. If I were to turn water in the field when too cold, would that keep the frost off, and if so, would I have to turn water down each row, or would one furrow full of water to about every fourth or sixth row be enough?

You can prevent frost by smudging for potatoes just as you can for other vegetables. The potato, however, needs little protection of this kind and will endure a light frost which would be destructive to tomatoes, melons, and other more tender growths. Unless you have a very frosty situation, you can certainly grow potatoes without frost protection, and they should be planted earlier than February first if the ground is in good condition. The great secret of success in growing potatoes in southern California is to get a good early start before the heat and drought come on. Water will protect from frost if the temperature only goes to about 28 degrees and does not stay there too long. The more water there is exposed the longer may be the protection, but probably not against a lower temperature.

Growing Sweet Potato Plants.

How shall I make a hot-bed to raise sweet potato plants? I don't mean to put glass over bed, but want full description of an up-to-date outfit for raising them.

Manure hot-beds have been largely abandoned for growing sweet potato slips, though, of course, you can grow them that way on a small scale or for experiment. In the large sweet potato districts, elaborate arrangements for bottom heat by circulation of hot water or steam are in use. In a smaller way hot air works well. The Arizona Experiment Station tells how a very good sweet potato hot-bed at little cost is constructed as follows: A frame of rough boards seven feet wide, twenty feet long and fourteen inches deep is laid down over two flues made by digging two trenches one foot deep and about two feet wide, lengthwise of the bed. These trenches are covered with plank or iron roofing, and are equipped with a fire pit at one end and short smokestack at the other.

Four inches of soil is filled into this bed and sweet potatoes placed upon it in a layer which is then covered with two or three inches more of soil. Large potatoes may be split and laid flat side down. The whole bed is then covered with muslin, operating on a roller by which to cover and uncover the bed. Thus prepared, the bed may easily be kept at a temperature of 60 to 70 degrees F. by smouldering wood fires in the fire boxes. The potatoes, kept moist at this temperature, sprout promptly and will be ready to transplant in about six weeks. A bed of the size mentioned will receive five to seven bushels of seed roots, which will make slips enough to plant an acre or more of potatoes.

Growing Sweet Potatoes.

Please inform me how to keep sweet potatoes for seed; also how many pounds it takes for one acre, and what distance apart to plant, and the time to plant.

Sweet potatoes may be kept from sprouting by storage in a cool, dry place. Sweet potatoes are not grown by direct cutting of the tuber as the ordinary potato is, but the tubers are put in January or later in a hot bed and the sprouts are taken off for planting when the ground becomes warm and all danger of frost is over in the locality. The number of sprouts required for an acre is from five to ten thousand, and a bushel of small sweet potatoes will produce about two thousand sprouts if properly handled in the hot bed, which consists in removing the sprouts when they have attained a height of five or six inches, and in this way the potatoes will be yielding sprouts in succession for some time. The sprouts are planted in rows far enough apart for horse cultivation. They are usually hilled up pretty well after starting to grow well. They cannot be planted until the danger of frost is over, for they are much more tender than Irish potatoes.

Sweet Potato Growing.

In planting sweet potatoes, do we have to make hotbeds just like those for tomatoes, or if just a plain seed-bed will do? Is it necessary to irrigate them or not?

You can bed your sweet potatoes in a warm place on the sunny side of a building or board fence, and get sprouts all right. You will, however, get them sooner and in greater numbers by using a slow hotbed in which the manure supply is not too large. The fact that sweet potato growers do use some artificial heat, either from manure or by piping bottom-heat in their propagating houses, is a demonstration that such recourse is desirable to get best results. The necessity of irrigation depends upon the soil and its natural moisture supply. On a fine retentive loam, the crop is chiefly made without irrigation, if the plants are all ready to put out in the field as soon as it is safe. If you are late in the planting, or if the soil is dry or likely to dry before the tubers are grown to good size, irrigation, some time ahead of the need of the plant, is essential.

Sweet Potatoes.

What kind of soil and climate does it take to grow sweet potatoes, and can I grow them in any part of Contra Costa county, and about what time is the best to plant them?

Sweet potatoes do best in a light warm loam which drains well and does not bake or crust by rain or irrigation. Sprout the tubers in a hot-bed or cold-frame in February and break off the shoots and plant as soon as you are out of danger by frost. Sweet potatoes are more tender than common potatoes. There are places in Contra Costa county where they do well, though some parts of the county do not have enough summer heat.

Sweet Potatoes Between Fruit Trees.

I am expecting to grow a fall crop of about twenty acres of sweet potatoes. The land is a heavy, sandy loam in the interior, which has been set out this spring to almonds, apricots and prunes. I wish to grow sweet potatoes between trees. Would an irrigation every forty days be often enough? Also, if either sweet or Irish potatoes grown between rows are harmful to either of the varieties of fruit mentioned?

We see no reason why you should not get your crop, providing you do not have to run the plants into the frosty period, and sweet potatoes will not, of course, stand frost as well as the common potato. The moisture which you propose to give ought to be enough for a retentive soil in connection with good cultivation until the vines cover the ground. Growing any crop between orchard trees is apt to be an injury to the trees, because of the spaces which are not and cannot be adequately cultivated, so that the ground around the trees is apt to become compacted either by the run of water or the lack of cultivation, or both. Our observation has been that Irish potatoes are no more injurious than other crops. Any crop will injure young trees if it takes moisture they ought to have or interferes with good cultivation of the land.

Giant Japanese Radish.

In discussing sakurajima (giant Japanese radish) Eastern publications advise planting late, about August 1, and not earlier than July 1. What can you tell me about the plant here?

The Asiatic winter radishes can be successfully planted in California in July or August if the soil is thoroughly saturated by irrigation before digging and planting. It is, however, not so necessary to begin early in California as at the East, because our winter temperatures favor the growth of the plant, while at the East they have to make an early start in order to get something well grown before the ground freezes. For the growth of winter radishes, then, in California you can wait until the ground is wet thoroughly by the rain, which may be expected during September, and afterward you can make later plantings for succession at any time you desire during the rainy season. This applies to all kinds of radishes.

Rhubarb Rotting.

I have planted rhubarb roots in the San Joaquin valley and find the root crowns rot below the surface.

The old-fashioned summer rhubarb usually goes off that way in very hot localities. If there is too much alkali or hardpan, or if planted too late, the same results will be had with any sort of rhubarb. Where it is very hot, plants, irrigated in the morning near the plants, scald at the crown and die in a few days. If irrigated in the afternoon and the ground worked before it gets hot the next day fine results are obtained. The winter rhubarb varieties do well in hot districts if the roots are planted from September 15 to May 1, while in cooler sections, April, May, June and July are the best months and will insure a crop the following winter.

Squashes Dislike Hardship.

What caused these squashes, of which I send you samples, to be so hard and woody? They were grown without irrigation.

Your squashes were grown without irrigation under conditions which were too dry for them and became inferior in quality. Possibly the variety itself is not of good quality or the specimen from which the seed was taken may have been inferior. A squash, in order to be tender and acceptable, needs rich feeding and plenty of drink. Otherwise, it is apt to resent ill treatment by very undesirable growth.

Harvesting Sunflowers.

What is the method used in saving or threshing the seed from the Giant

Russian sunflower?

Cut off the seed heads of your sunflowers when the seed seems to be well matured but before any of it falls away from the head. Throw these heads on a smooth piece of ground or a tight floor and when they become thoroughly dry thresh out the seed with a flail, removing the coarse stuff with a rake and afterwards cleaning the seed by shoveling it into the wind so that the light stuff may be blown away. A more perfect cleaning afterwards could be secured with a grain fanning mill or a simple sieve of the right mesh.

Irrigating Tomatoes.

How much water does it take (in gallons or cubic feet) to properly irrigate an acre of land for tomatoes? The soil is adobe, and the customary way of planting tomatoes is 6 feet apart each way, plowing a trench of one furrow with the slope of the land for irrigating, that is, a trench between every row and a cross trench as a feeder. The land is low and in the driest part of the year the surface water is from 2 to 3 feet beneath the top of the ground.

It is not possible to state a specific quantity of water for any crop, because the amount depends to such a large extent upon the retentiveness of the soil, the rate of evaporation and the kind of cultivation. The best source of information is the behavior of the plant itself, bearing in mind that tomato plants require constant but not excessive moisture supply, and that if moisture is applied in excess it will promote an excessive growth of the plant, which will cause it to drop its blossoms and therefore be unsatisfactory and unproductive. In such land as you describe no irrigation whatever would be desirable except in years of short rainfall, and such land, if properly cultivated, would always furnish moisture enough by capillary action to support the growth of the plant.

Less Water and More Heat.

What chemicals should I put into the soil to insure a good crop of vegetables, such as tomatoes, string beans, or other over-ground producers? Last year my tomatoes and string beans grew plentifully, but never produced any tomatoes or beans, yet turnips and parsnips were all right.

Vegetables which behave like your tomatoes and string beans, making too much growth and not enough fruit, do not need fertilization. The land is perhaps too rich already, or you may have used too much water. Use less water so that the plants will make a more moderate growth, and they will be fruitful if the season is warm enough in the later part of summer. This, of course, would be one of the drawbacks to growing tomatoes and beans in San Francisco. Turnips and parsnips do well with less heat. You may have to modify the San Francisco summer climate by wind screens or glass covers.

Continuous Cropping With the Same Plant.

What would happen on the crops of cucumbers, tomatoes and eggplants, etc., planted on the same place continuously?

There would be in time a decadence of crop from soil exhaustion, but that you could prevent by fertilization. The greatest danger from continuously growing these vegetables on the same land is the multiplication of bacteria which injuriously affect them, in the soil. The plants which you mention are all subject to "wilt" diseases from this cause, therefore, they should have new ground. If you have to use the same garden ground continuously, the plants which you mention should be rotated with root crops or with other kinds of vegetables, so as to frequently change plants and soil within the general area which has to be used for them.

Big Worms on Tomatoes.

I have a nice patch of tomatoes in my garden, and only recently I notice large green worms on them with one large brown horn on their head. They strip the leaves off. They look to me like a tobacco worm.

They are tobacco worms; that is, they are the larvae of hawk moths, some of which take tobacco, tomatoes, grapevines and many other plants, including some of the native weeds of your valley. Pick them off and crush them, or give them a little snip with the scissors if you do not like to handle them. They are so large and easily found that such treatment is easily applied, as in "worming tobacco."

Loss of Tomato Bloom.

I have tomato plants which are very strong and healthy and full of blossoms, but there is something cutting the blossoms off and just about to ruin my plants.

The trouble with your tomato plants is that life is too easy for them, that they have so much moisture and plant food that they can grow comfortably and rapidly without thought of the future. So, because they do not have to think of making fruit, the blossoms drop off. This is a very common occurrence with tomatoes, especially in home gardens where the owners have not the experience or the information on the subject that they might have, and give the tomatoes too much water. Many other plants act the same way and will not set fruit while they can grow easily, and only begin to produce when they have made a great growth or when moisture begins to get a little short. If you irrigate the tomatoes, stop, and put no more water on until the plant begins to set fruit as if it meant business, or gives some sign that water would be appreciated. If the ground is naturally moist you will have to wait until the plants make more growth and the weather gets drier and hotter, and the plants will then set fruit. Some growers have found that by trimming up the vine and staking it, the fruit sets much more readily.

Part III. Grains and Forage Crops

Wants Us to Do the Whole Thing.

Can you help, me to determine a good product to plant so-mewhere in California; also what particular section would be most suitable for the raising of that which you would advise? I wish a crop of permanent nature (as orchard trees). I also desire advice on some product which would give a quick return while I am waiting on the more permanent one to mature and bear. I have not procured land yet, and am thinking seriously of trying to get government land, therefore, you are free to give me the best location for the raising of that which you would, suggest. I want a money-making product and one which is not already overdone.

The choice of crops depends quite as much upon the market demand and opportunity as it does upon the suitability of the soil and local climate. Choice of crops indeed involves almost the whole business of farming, and although we can sometimes give a man useful suggestions as to the growth of plants and the protection of plants from enemies, we cannot undertake to plan his farming business for him. He must form his own opinions as to what will be most marketable, and therefore profitable, if he succeeds in getting a good article for sale. A wise man at the East once said: "You can advise a man to do almost anything. You can even select a wife for him, but never commit the indiscretion of advising him what to grow to make money. That is a matter he has to determine for him-self."

Pasturing Young Grain.

Would it be advisable to herd milch cows for a few hours each day on a field of black oats which is to be grown for hay? The oats are now about four inches high and rank, as the land was pastured last year. The land is sandy, rolling soil and will soon be dry enough so that the cows would not injure the plants. The idea is that the leaves which are green now will all dry up and are really not the growth which is cut for hay; therefore, I should think it would do no harm to feed it down a bit.

Over-rank grain with abundant moisture will make a more stocky growth and stand against lodging if pastured or mowed. The leaves which you speak of as being lost in the later growth of the plant serve an important purpose in making that growth, and removing them is a repressive process which is not desirable when rain is short. We should allow the plants to push along into as good a growth of hay as a dry year's moisture will give.

Dry Plowing for Grain.

We have land that we could very easily plow now with our traction engine and improved plows, but the people here claim that it does not pay to dry-plow, that is, before the land has had a good rain on it and the vegetation has started. I believe in dry plowing. Two of our oldest farmers in Merced county dry-plowed, that is, they commenced plowing as soon as harvesting was over.

If the rainfall is small and likely to come in light showers, dry plowing, if it turns up the land in large clods, might yield poorer results than land which is plowed after rain, because there would be so much moisture lost by drying out from the coarse surface when it came in amounts not adequate for deep penetration. Plowing after the rain for the purpose of killing out the foul stuff which starts is, however, quite another consideration. It is a fact that dry plowing and sowing is not now desirable in some places where it was formerly accepted, because the land has become so foul as to give a rank growth of weeds which choke out the grain at its beginning.

174

Such land can be cleaned by one or two shallow plowings and cultivations after there is moisture enough to start the weeds to growing. These are local questions which you will have to settle by observation. In a general way, it is true that opening the surface of the ground before the rains, reduces the run-off and loss of moisture, but whether there would be any loss of moisture by run-off or not depends upon the slope of the land and also upon the way in which the rain comes, and the total amount of moisture which is available for the season.

Sub-varieties of California Barley.

Can you tell where I can buy seed of varieties of California six-rowed barley, described as "pallidum" and "coerulescens," and what the seed will cost?

No one knows where the six-rowed barley, known as "common" barley in this State, came from, nor when it came. It has been here since the early days and it has naturally shown a disposition to vary, so that it is quite possible to select a number of types from any large field, of it. These variations have been studied to some extent by Eastern students who are endeavoring to develop American types of barley for brewing purposes as likely to be better than the brewing varieties which are famous in Europe. In Europe brewing barleys are chiefly two-rowed. Under California conditions the plant is able to develop just as good brewing grains on a six-rowed basis, and this seems to be a commendable trait in the way of multiplying the product. The names "pallidum" and "coerulescens" indicate two of these varieties recognized by Eastern students. It is not possible at this time to get even a pound of selected grain true to this type, and no one knows when it will be worked out to available quantities.

Chevalier Barley.

Has Chevalier barley more value to feed hens for egg production than common feed barley or wheat?

Chevalier barley is no better for chicken feed than any other barley which is equally large and plump. Brewers like Chevalier because of its fullness of starch to support the malting process; also, because it is bright, that is, white, and not stained or tinged with bluish or reddish colors. Color points do not count for chicken feed, but good plump kernels do. Besides this, however, darker kernel (not chaff) usually indicates more protein, and therefore a darker kernel of either wheat or barley might be more valuable for feeding. A hard, horny kernel is richer than a softer, more starchy one, either in wheat or barley.

Barley on Moist Land.

What would you do with land subject to overflow by the Sacramento when that river rises 20 feet, and which you wanted to plant to barley this season? Would you take a chance on the river rising that high this year, or wait until after that danger was over, and take a chance on not getting enough rain to make the grain come up; also, if the river did come up for 48 hours after the grain was in, but did not wash, would the grain be lost? Should the grain be planted deeper than on ordinary land, and, if so, should a drill be used? How much seed should be sown per acre on good river-bottom soil?

Get the barley in and watch for the overflow rather than to fear it. An overflow for 48 hours would give you the greatest crop you ever saw, unless it should be in a settling basin and the water forced to escape by evaporation. From your description we judge that this is not so and that the land clears itself quickly from an overflow.

Depth of sowing depends upon the character and condition of the soil - the lighter and drier the deeper. By all means use a drill if the soil is dry on the surface. Short rainfall makes the advantage of drill seeding most conspicuous. On the University Farm 22 trials gave an average gain of over 10 per cent in yield. The difference would be much greater in a dry year; it might be 25 per cent greater, possibly, and save high-priced seed at the same time, as about 90 pounds of seed per acre will do, instead of 120 pounds broadcast, in accordance with the approved heavy seeding practice on the river lands.

Barley and Alfalfa.

I have some alfalfa which is a poor stand. Can I disc it up heavily and seed in some barley for winter pasture?

You can get barley into your alfalfa as you propose, but you should not seed until fall. The more barley you get into your alfalfa, however, the less alfalfa you will have afterward. If you want to improve your afalfa, keep everything else out of the field and help the plants by regular irrigations during the balance of the growing season.

Beets and Potatoes.

Which is the best for dairy cows, plain red mangels or a cross between these and sugar beets? Can you suggest a more profitable variety of potato than the Oregon Burbank?

If you can get a cross which gives you more tonnage than a mangel and a higher nutritive content you would have something better to grow. The first point you have to determine by growing the two side by side and weighing the product; the nutritive value of each will have to be determined by chemical analysis. Until these determinations are actually made a comparison of desirability is nothing but conjecture. There are several other potatoes which are sometimes more profitable here and there for early crop when grown in an early locality. If you are not in an early locality you are obliged to produce for the main crop, and nothing, to our knowledge, sells as well as the Burbank, if you get a good one.

Beets for Stock.

Will sugar beets grow on black alkali land? How many pounds of seed per acre should be used and when is it time for sowing in the San Joaquin valley? Which kind would be best for cows?

Beets will do more on alkali than some other plants, but too much alkali will knock them out. You must try and see whether you have too much alkali or not. You can sow at various times during the rainy season, for the beets will stand some frost. Sow 8 pounds per acre in drills 2 1/2 to 3 feet apart, so as to use a horse cultivator. For stock you had better grow large stock beets like marigolds or tankards - not sugar beets. It costs too much to get sugar beets out of the ground, because it is their habit to grow small and bury themselves for the sake of the sugar maker, while stock beets grow largely above ground.

Summer Start of Stock Beets.

How can I make Mangel Wurzels grow in hot weather? The land is level and can be irrigated by flooding or ditching between the rows. How often should the water be applied, and which method used? The land is in fine shape; a sandy loam bordering on to heavier land.

Wet the land thoroughly; plow and harrow and drill in the seed in rows about 2 1/2 feet apart. This ought to give moisture enough to start the seed. Cultivate as soon as you can see the rows well. Irrigate in a furrow between the rows about once a month; cultivate after each irrigation.

Corn Growing for Silage.

With fair cultivation, will an acre produce about 10 tons of ensilage without fertilization - it being bottom land? How should it be planted? - the rows closer together than 3 feet, or should it be planted the usual width between rows, and thick in the rows? If fertilizers were to be used, what kind would you recommend? Would you recommend deep plowing followed by a packer and harrow so as to preserve the moisture?

You ought to be able to get 10 tons of silage per acre from corn grown on good corn land. It can be best grown in rows sufficiently distant for cultivation, closer in the row than would be desirable for corn, and yet not too crowded, because corn for silage should develop good ears and should be cut for silage about the time when the glazing begins to appear. If your land needs fertilization, stable manure or a "complete fertilizer" of the dealers would be the proper thing to use. It would be very desirable to plow corn land deeply the preceding fall, followed by a packer or harrow to settle down the land below, but do not work down fine. Keep the surface stirred from time to time during the winter and put in the crop with the usual cultivation in the spring as soon as the frost danger is over.

Irrigation for Corn.

What amount of water is necessary per acre for the best possible yield of corn under acreage conditions and proper cultivation in the San Joaquin or Sacramento valleys?

No one can answer such a question with anything more than a guess. It depends upon how much rain has fallen the previous winter, how retentive the soil is naturally, and what has been done to help the soil to hold it. Nearly all the corn that is grown is carried without any irrigation at all on moist lowlands, which may be too wet for winter crops. If you demand a guess, make it six acre-inches, with a good surface pulverizing after each run of water in furrows between the rows. This water would be best used in two or three applications.

Eastern Seed Corn for California.

The question has been raised as to Eastern-grown seed corn, comparing it with California-grown seed. Some claim that the former does not yield well the first season.

We cannot give a complete refutation of the impression that Eastern seed corn does not yield well the first season in California. It is a somewhat prevalent impression. All that we can announce now is that we have grown collections of Eastern seed corn and have found the product quite as good as could have been expected, and did not encounter, apparently, the trouble of which you write.

Need of Corn Suckering.

To insure the best crop of corn possible, does it pay to sucker it or not?

The removal of suckers is a matter of local conditions largely in California, and growers are getting out of the habit of suckering. In some places suckering is needed, and in others it apparently does not pay to do so, although with very rare exceptions a larger yield can be secured by suckering than without.

Cow Peas Not Preparatory for Corn.

What time of the year can cow peas be planted, and can the entire crop be plowed under in time for planting field corn?

Cowpeas are very subject to frost. They are really beans, and therefore can be grown in the winter time only in a few practically frostless places. Wherever frosts are likely to occur they must be planted, like beans and corn, when the frost danger is over. Field peas, Canadian peas and vetches are hardy against frost and therefore safer for winter growth, and treated as you propose they may be preparatory for corn-growing providing you plow them under soon enough to get a month or more for decay before planting the corn.

Oats and Rust

Is there any variety of oats that is rust-proof, or any method of treating oats that will render them rust resistant? We are situated on a mountain, only about 12 miles from the coast, and have considerable foggy weather, which most of the farmers here say is the cause of the rust.

There is no way of treating oats which will prevent smut, if the variety is liable to it. There is a great difference in the resistance of different varieties. A few dark-colored oats are practically rust-proof, and you can get seed of them from the seedsmen in San Francisco and Los Angeles. Such varieties are chiefly grown on the southern coast. Foggy weather has much to do with the rust, because it causes atmospheric moisture which is favorable to the growth of the fungus, which is usually checked by dry heat, and yet there are atmospheric conditions occasionally which favor the rust even in the driest parts of the State. The fog favors rust, but does not cause it. The cause is a fungus, long ago thoroughly understood and named puccinia graminis.

Midsummer Hay Sowing.

Can I sow oats or barley in July upon irrigated mesa land, with the object of making hay in the fall? Which of the two would do the better in summer time? I have plenty of water.

We have never seen this done to advantage. If you desire to try it, irrigate thoroughly and plow and sow afterward. Use barley rather than oats and irrigate when the plant shades the land well, if you get growth enough to warrant it. It will be easier to get the crop than to figure a profit in it.

Loose Hay by Measure.

How many cubic feet should be allowed for a ton of alfalfa hay loaded on a wagon from the shock? I must sell more or less in that way, as no scales are near enough to be used.

It is a proposition, as to the weight of loose hay, which could of course keep changing the higher you built the load on the wagon. It is easier to give figures on weight from a stack in which there has been something like uniform pressure for a time. In the case from a

30-day stack it is common to allow an eight-foot cube to a ton, etc. Perhaps you can guess from that.

When to Cut Oat Hay.

To make the best red oat hay should it be cut when in the "milk," "dough" or nearly ripe!

It should be cut in the "soft dough" or, as some express it, "between the milk and the dough." This is probably as near an approach in words as can be made to that condition which loses neither by immaturity or by over-maturity from the point of view of hay which is to get as much as can be in the head without losing nutritiveness in the straw. Of course there are other conditions intruding sometimes, like the outbreak of rust or the premature ripening through drought. In such cases care must be taken not to let the plant stand too long for the sake of reaching an ideal condition in the head - which for lack of favorable growing conditions the plant may not be able to reach.

Rye for Hay.

When is the best time to cut rye for hay, and how should it best be handled? Would it be well to cut it up and blow it into the barn, and would it do all right for silage?

Rye makes poor hay on account of its woody stems and must be cut earlier than other grains. After that it is handled as is other hay. Cutting it up would probably be more of a help than to other grain hay. It could be put into the silo, but would of course have to be cut pretty green and would have to run through a cutter and blower. Putting it in whole would be out of the question. In the silo, the fermentation would largely overcome the woodiness of the stems. It would also as a silage balance up nicely with alfalfa, and the best way to do would be to mix it with alfalfa when putting it in.

Rye in California.

Which kind of rye is the hardiest, the best yielding, and the best hay varieties in your State?

Rye is the least grown of all the cereals in California, and no attention has been paid to selection of varieties. That which is produced is "just rye," of some common variety which came to the State years ago and still remains. No rye is grown for hay, as the toughness of the stem renders it undesirable for that purpose. There is a certain amount of rye grown for winter feeding. This is grown in the foothills principally and it serves an excellent purpose, but it is fed off before approaching maturity.

That Old Seven-Headed Wheat.

We are sending you some heads of grain which was grown in this county. The land was planted with an imported Australian wheat, which we believe the smaller heads to be, but the wheat is about evenly mired with grain like the large heads, which we think to be a species of barley.

The grain is an old, coarse, bearded wheat which is continually appearing in fields of ordinary grain and naturally excites interest among all to whom the variety is a novelty. It is the old seven-headed Egyptian wheat, which has never proved of any cultural value, because its manifolding of the head is of no advantage. It is better to have a straight well-filled head than to have a branching

184

head of this kind. This matter has been fully demonstrated by experience during the last thirty or forty years, not only in this State, but in other States, for the variety has a way of getting around the world, and seed has sometimes been sold at exorbitant prices to people who have been persuaded that it is of particular value.

Speltz.

I have heard of a Russian grain called "Speltz" or "Emmer." Can I raise it successfully and, if so, what is the very best time of year to sow some for the best crop obtainable? Can it be sown in the fall, say November? Would springtime be a better time to sow it on soil that is very soft in winter?

If your land yields good crops of wheat or barley or oats, you have little to expect from speltz or emmer. This is a grain generally considered inferior to those just mentioned and advocated for conditions under which the better known grains do not do well. It is hardy against drought and frost, particularly the latter, and is, therefore, chiefly grown in the extreme north of Europe. It may be sown in the fall or in the spring in places where rains are late and carry the plant to maturity.

Italian Rye Grass.

What kind of grass is enclosed? Also the best method to eradicate it?

The grass is the Italian rye grass, or as it is sometimes called, the Italian variety of the perennial rye grass. It is proving a very satisfactory grass in California for moderate drought resistance and for winter growing, and a great deal of it is being sown for these purposes. You can readily kill it out by cultivation, but most people are more occupied with its propagation than with its destruction.

Fall Feed.

Can I irrigate and plant a forage crop n July to feed dairy cows this fall and winter? Would you recommend cow peas or some kind of sugar corn? If cow peas, how many pounds to the acre?

If you wet down the land thoroughly and then plow and harrow and plant either cow peas or Indian corn, you ought to get a good green crop before frost. Drill in or drop the seed in rows about three feet apart and keep cultivating and irrigating as long as you can get through without injuring the crop too much. Use about 40 pounds of cow peas to the acre.

Hurry-up Pasture.

What can I plant this fall which would produce pasturage for a small amount of stock this winter, and until I can get the land under irrigation and seeded to alfalfa?

For quick fall and winter growth nothing is better probably than oats and vetches sown together as soon as you get rain enough to plow, but it would be a question whether it is worth while to work for that, because you ought to get your land ready for February sowing of alfalfa and that will keep the land busy after the rain gets it into working condition.

Johnson Grass.

I am informed that Johnson grass makes fine hay. I have not sown the seed yet, but would like to know if the hay is good and if it will grow on dry land. I have the seed on hand, but do not want to sow it if it is not good.

Johnson grass is poor, coarse stuff. The plant is most valuable for grazing when young. Johnson grass will not grow on really dry land, but it will take the best moist land it can find and hold on to it. It is sensitive to frost and is not a winter grower except in the absence of frost.

Improving Heavy Land for Alfalfa.

My land is very heavy, red loam, and crusts over very hard in dry seasons. I would like to know if it would be best to use barnyard compost over the surface as a mulch, or would it be best to use plain straw for that purpose?

A very heavy soil can be brought into better surface condition for alfalfa by plowing in stable manure as soon as possible after the fall rains, in order that the manure may have opportunity to become disintegrated and mixed with the soil by the time for alfalfa sowing, which is from February to April - whenever the heavy frosts of the locality are over. For a small piece, you might get a better stand by using a light mulch of disintegrated coarse manure or even straw, scattering it after the sowing, but for a large acreage this would involve too much labor. It is not desirable to work in much manure or other coarse stuff at the time of sowing the seed, but you can make a light surface application after the plant has made a start.

Cultivating Alfalfa.

When is the best time to cultivate alfalfa, and how often during the season is it advantageous to do so? Which is the best implement to use?

Cultivated alfalfa is a term applied to alfalfa sown in rows and allowed to grow in narrow bands with cultivated land between, and the irrigation is then done in a furrow in the narrow cultivated strip. This will give thriftier growth and perhaps more hay to the acre than flooded, broad-casted alfalfa, but it will cost so much more that the acre profit would probably be less. This is an intensive culture of alfalfa, which is still to be tested out in California, if any one should be inclined to do it. Some one-cow suburbanite would be in condition to try the scheme first. Probably you refer to disking, and for that an ordinary disk is used with the disks set pretty straight to reduce the side cutting, and this is done at different times of the year by different growers. By doing it when the ground gets dry in the early spring much of the foul stuff is cut out before the alfalfa starts strongly. But disking seems to be good whenever in the year the soil is dry enough to take it well.

Suburban Alfalfa Patch.

How can we rid the alfalfa of weeds? As we are obliged to hire help, and do not succeed in getting the hay cared for until we have mostly stalks without leaves, I have put the cow on it to pasture it off.

The cow knows how to handle it, but you will not get as much alfalfa as if you cut and carried it to her. If you cut sooner you will get rid of many plants which are propagated by the seeds which they produce, and you will also get better hay, more leaves and fewer stalks. Cut it about the time it begins to bloom, not waiting for the full bloom to appear.

Alfalfa and Bermuda.

I have land which was seeded to alfalfa some 15 years ago and has been pastured continuously until it was almost all Bermuda. I had it thoroughly plowed, disk harrowed and sowed to oats; disk harrowed in, and drag harrowed. After cutting for hay this year I intend putting it in Egyptian corn in rows, so it can be cultivated to get rid of Bermuda. I have also been advised to plow the land immediately after harvesting corn and let it lie until next January and then plow and sow to barley and alfalfa as I wish to grow alfalfa. Kindly let me know if method is right. The land is sandy loam and under irrigation.

Whether you will fully succeed against Bermuda grass or not is doubtful. It is probable, however, that you can reduce the Bermuda so that other cultivated crops can be continuously grown. Common experience is that Bermuda will hold on unless you have hard freezing of the ground to a considerable depth, as they have in the northern States. The best use that you can make of land infested with Bermuda is to get as good a stand as you can of alfalfa and let the alfalfa fight for itself. The combination of alfalfa and Bermuda grass makes very good hay or pasturage. We should, however, sow the alfalfa alone and not handicap it by sowing with barley. The Bermuda will smile at that advice. Egyptian corn can be planted in rows, 2 1/2 to 3 feet between the rows to admit of easy cultivation

Bermuda Grass.

What is the value of Bermuda grass as a forage crop for cattle, more particularly dairy cows?

Bermuda grass is generally condemned because of getting in places where it is not desirable and of being almost impossible of eradication therefrom. Still, Bermuda grass will make good pasturage on land which is too alkaline to make other crops, and therefore is highly esteemed by some owners of waste lands in the San Joaquin valley. It is good pasturage and is most easily propagated by cutting the roots up into short pieces by use of the hay-cutter, nearly all the pieces retaining an eye which will make a new plant. It is easy to get in and hard to get out.

Salt Grass and Alfalfa.

I have some land in Sutter county and it has some of this salt grass in spots. I am about to take a twenty-acre piece and put in alfalfa, but some old-timers tell me that the salt grass on it is bad stuff to handle.

Your trouble will probably be not so much the salt grass, but the alkali in the soil which the salt grass can tolerate and which other plants cannot stand. You cannot then substitute alfalfa for salt grass without getting the alkali out of the soil, and you cannot do this without having sufficient drainage so that the rainfall may wash the alkali out from the soil and carry it away in the drainage water. You probably cannot get a satisfactory growth of alfalfa on the spots where the salt grass has established itself, although the land round about may be very satisfactory to alfalfa.

Giant Spurry.

I would like information about spurry. How much frost will it stand?
What is time for sowing? Its value as crop to plow under?

From a California point of view, spurry is a winter-growing weed which has been approved by orchardists in Sonoma county because it yields a considerable amount of vegetation for turning under with the spring plowing of the orchard. For this purpose it should be sown at the beginning of the rainy season. Its value as a crop to turn under depends upon the amount of growth you can get. It is not a legume and, therefore, does not have the value of the nitrogen-gathering plant. Still, it yields humus and, therefore, is valuable for winter growing as ordinary weeds, grasses, grains, etc., are.

Light Soil and Scant Moisture.

Advise me as to plowing under a crop of last year's weeds where I intend to plant beans, corn, etc. The soil is "slickens," on the Yuba river, and the weeds grew up last year in a crop of volunteer barley, which was hogged off. I expect to plow five inches deep, and calculate that the barley straw and weeds will contribute to the supply of humus, which is always deficient in most of our soils. I expect to try to grow beans without irrigation, and wonder if the trash would hold the soil too open so as to dry them out.

Considering the character of the soil which you describe and the shallow plowing you intend we should certainly burn off all the trash upon the land. With deep plowing early in the season this coarse stuff could be covered in to advantage, but it would be dangerous to do it in the spring. Clean land and thorough cultivation to save moisture enough for summer's growth is the only rational spring treatment.

Clovers and Drought.

I have sandy loam with some alkali. In wet years it is regarded as too damp in some places. Can you give me any information on the following points? I have practically no water for irrigation and I feel sure that alfalfa would not grow without it. Do you think that clover would make one or more cuttings without water?

Red and white clover are less tolerant of drought than alfalfa, which, being a deep-rooting plant, is especially commended in dry-farming undertakings. Red clover will grow better on low wet lands than will alfalfa, but the land must not dry out or the red clover will die during the dry season. None of the plants will stand much alkali.

Clover for Wet Lands.

What kind of alfalfa will do best on sub-irrigated land which is very wet? I have sown it in alfalfa and it grows finely for two or three years, but then the roots rot and die.

It is impossible to make any kind of alfalfa grow well on very wet land, that is, where the water comes too near the surface. Alfalfa has a deep-running tap root which is very subject to standing water. You can get very good results from the Eastern red clover on such land, because the red clover has a fibrous root which is content to live in a shallow layer of soil above water. But red clover will not stand drought as well as alfalfa, because it is shallower rooting. It is necessary, therefore, that water should be permanently near the surface or surface irrigation be frequently applied, in order to secure satisfactory growth of red clover in the drier sections of California. It is also necessary that neither land nor water carry alkali.

Frosted Grain for Hay.

The freeze struck us pretty severely. I had 125 acres of summer-fallowed wheat which I had estimated to make 20 sacks to the acre of grain. It was breast high in places already, and was just heading out. The frost pinched the stalks of this grain in several places and the heads are now turning white. It is ruined for grain. There is lots of fodder in it, and it should be made into hay. If so, should it not be cut and cured at once? What is the relative worth of such hay as compared with more matured hay? Would the fact that it is frozen make it injurious to feed?

If the whole plant seems to be getting white, the sooner it is cut the better. If the head is affected and the leaf growth continued, cutting might be deferred for the purpose of getting more of it. Hay made from such material will not be in any way dangerous, although it would be inferior as containing less nutritive and more non-nutritive matter. Such hay would seem to be most serviceable as roughage for cows or steers in connection with alfalfa hay or some other feed which would supply this deficiency.

Forage Plants in the Foothills.

We have 3,000 acres of foothill land and hope to be able to irrigate some land this spring and wish to know the best forage crops, for sheep and hogs, especially. Kafir corn, stock peas, rape, sugar-beets and artichokes are the varieties about which we desire information.

Where you have irrigation water available in the foothills you can get a very satisfactory growth of red clover. We have seen it doing very well on sloping land in your county where water was allowed to spill over from a ditch on the ridge to moisten the slope below. Winter rye and other hardy stock feeds could also be grown in the winter time on the protected slopes with the rainfall. Some such plants are not good summer growers, owing to the drought. Rape is a good winter grower by rainfall, but not so satisfactory as vetches and kale. Sugar beets are not so good for stock purposes as stock beets, which give you much more growth for the same labor and are more easily gathered because they grow a good part out of the ground. They will stand considerable freezing and may be sown at different times throughout the year, whenever the land is moist, either by irrigation or rainfall. Artichokes are of doubtful value. We have never found anyone who continued to grow them long. Of course, on good, deep land, with irrigation, nothing can be better than alfalfa as supplementary to hill range during the summer season.

Winter Forage.

At what time of the year should I plant kale, Swiss chard, etc., so as to have them ready for use during the months from February to June?

You should plant Swiss chard, kale, etc., as soon as the ground is sufficiently moist from the rain in the fall. In fact, it would be desirable for you to plant the seed earlier in boxes and thus secure plants for planting out when the ground is sufficiently moist. These plants are quite hardy against frost, and in order to have them available by February, a start in the autumn is essential.

A Summer Hay Crop.

What can I put on the land after the oat crop is taken off to furnish hay for horses during the coming winter? I had thought millet would be good. I have water for irrigation.

You could get most out of the land you mention during the hot season by growing Kafir corn or milo, cutting for hay before the plant gets too far advanced. If your land can be flooded and takes water well, so that you can wet it deeply before plowing, the sorghum seed can be broadcast and the crop cut with the mower while the stalks are not more than half an inch in diameter. This makes a good coarse hay. If you have not water enough or the land does not lie right for flooding, you can grow the sorghum in drills and irrigate by the furrow method, being careful, however, not to let the crop go too far if you desire to feed it as hay.

Teosinte.

What about "Teosinte," its food value, method of culture, and adaptability to our climate, character of soil required?

Teosinte is a corn-like plant of much lower growth than Indian corn. It may be of value as a forage plant on low, moist, interior lands in the summer season. It is very sensitive to frost and is, therefore, not a winter grower. It abhors drought and, therefore, is not a plant for plains or hillsides. It was grown to some extent in California 25 years ago and abandoned as worthless so far as tried.

Bermuda Objectionable.

Bermuda grass as pasture for summer to supplement burr clover and alfilaria in winter on the cheap hill pasture lands along the coast or the foothill ranges of the Sierras. Stock like it and do well on it, and I have noticed it growing in places where it had no water but the little rains of winter in southern California. So the question occurred to me, why should it not be a profitable pasture for the dry summers on the coast or foothill ranges of the State?

Bermuda grass will not make summer growth enough on dry pasture land to make it worth having. It will not make much growth in the rainy season because of frost, and if it has possession of the ground it will not allow either burr clover or alfilaria to make such winter growth as they will on clean land. Besides, this grass is generally counted a nuisance, because it will get into all the good cultivated land and it is almost impossible of eradication. Bermuda grass is of some account on alkali land where it finds moisture enough for free growth. We would not plant it in any other situation.

Rye Grasses Better than Brome.

I see in an Eastern seed catalogue "Bromus Inermis" very highly spoken of as pasturage. Do you know anything of it, and do you think it would be suitable for reclaimed tule land in the bay section?

Both English and Italian rye grasses have proved better than Bromus Inermis on such land as you mention. The latter is commonly known as Hungarian brome grass or awniess brome grass and it was introduced to this State from Europe about 25 years ago and the seed distributed by the University Experiment Station. Hungarian brome may be better on rather dry lands, although it will not live through the summer on very dry lands in this State, but we would rather trust the rye grasses or reclaimed lands, providing, of course, that they are sufficiently free from salt to carry tame grass at all. On the upper coast Hungarian brome has been favorably reported as an early-winter growing grass with comparatively low nutritive value, but is especially valuable because it will grow in poor soil. It is especially suited to sandy pasture and meadow lands and is quite resistant to drought. It is a perennial grass, reproducing by a stout rootstock, which makes it somewhat difficult to eradicate when it is not desired. It is desirable to keep stock off the fields during the first year to get a good stand.

Black Medic.

Will you kindly name the enclosed; also explain its value as forage!

The plant is black medic. It has been very widely distributed over the State during the last few years. It is sometimes called a new burr clover, which it somewhat resembles. It is not very freely eaten by stock and is apparently inferior to burr clover for forage purposes. It is a good plant to plow under for green manure.

Crimson Clover.

About crimson clover in California. Has it proved satisfactory? If so, can you give me data how to plant, etc.!

Crimson clover must be sown after frost, for it is tender. It will give a great show in June and July on low moist land. It is not good against either frost or drought. It has been amply tried in California and proved on the whole of little account.

California Winter Pastures.

We have a great deal of pasture land on which the native grasses yield less feed each year. A great part of this land can be cleared of brush and stone, ready for the plow, but what can we sow to take the place of the native pasture? The ground in many places is not level enough for alfalfa and in some places water is not available. Can we break up the land and sow pasture grasses as the farmers are exhorted to do at the East? The annual rainfall is from 12 to 15 inches.

The perennial grasses which they rely upon for pasturage in the East and which will maintain themselves from year to year, will not live at all on the dry lands of California, nor has investigation of the last twenty-five or thirty years found anything better for these California uplands than the winter growth of plants which are native to them. Such lands should be better treated, first by not being overstocked; second, by taking off cattle at the time the native plant needs to make seed, because, as they are not perennial, they are dependent upon each year's seed. After the plants have seeded, the land can be pastured for dry feed without losing the seed.

Of course, if one has land capable of irrigation he can grow forage plants, even the grasses which grow in moist climates, like the rye grasses, the brome grasses and the oat grasses, etc., which will do well if given a little moisture, but it will be a loss of money to break up the dryer lands with the idea of establishing perennial grasses upon them without irrigation. California pastures are naturally good. In early days they were wonderful, but they are restricted to growth during the rainy season, or for a little time after that, and are therefore suited for winter and spring pasturage, while the summer feeding of stock, aside from dry feed, should be provided from other lands where water can be used. The improvement of these wild pastures consists in a more intelligent policy for their production and preservation rather than an effort to improve them by the introduction of new plants. Pastures may, however, be often improved by clearing off the brush and harrowing in seed of burr clover, alfilaria, etc., at the beginning of the rainy season.

Alfilaria and Winter Pasturage.

Will alfilaria (Erodium cicutarium) grow well on the hills of Sonoma county partially covered with shrubs? I want something that will be food for stock another year. I have heard of alfilaria and that it grows well without being irrigated.

Alfilaria is a good winter-growing forage plant in places where it accepts the situation. It is an annual and therefore does not make permanent pasturage except where it may re-seed itself. On the coming of the dry season it will speedily form seed and disappear. It is therefore of no summer use under the conditions which you describe, nor is it possible to secure any perennial grass which will be satisfactory on dry hillsides without irrigation. Improved winter pasturage can be secured by scattering seed of common rye at the beginning of the rainy season, or of burr clover, both of which are winter-growing plants. Pasturage is also capable of improvement by being careful not to overstock the land, so that the native annuals may be able to produce seed and provide for their own succession. The secret of successful pasturage on dry uplands is to improve the winter growth. It is too much to expect much of them for summer growth without irrigation.

Grasses for Bank-Holding.

We desire a grass to be used on levees, to keep from washing. Bermuda or Johnson gross are dangerous to farming lands. What we desire is a grass that will grow in good dirt with no water to support it during most of the year, except the annual rainfall of Fresno county. Of course, this grass will also have to endure a great deal of water during the flooded season of the year. We have heard that the Italian rye grass would be suitable.

The rye grasses do not have running roots; therefore are not calculated to bind soil particles together as Bermuda grass does. If you want a binding grass, you must take the chances of its spreading to adjacent lands. Of course, if you could get a sod of rye grass it would prevent surface washing from overflow, etc., to a certain extent. We are not sure how far it would prevent bank cutting by the flowing water, for it makes a bunchy and not a sod-like growth. It would not live through the summer unless the levee soil keeps somewhat moist. The only way to determine whether you can get a permanent growth of it, will be by making a trial. Seed should be sown as soon as the ground becomes moistened by rain. It is a very safe proposition, because if it is willing to live through the summer, it is one of the best pasturage grasses for places in California where it will consent to grow, and it is not liable to become an annoyance by taking possession of adjacent land, because it would be readily killed by cultivation.

Alfalfa and Alkali.

I sowed several acres of alfalfa seed with a disc this season and none of it has come up. I think the reason for it not coming up is that the disc put it into the ground too deep. We sowed some by hand and it came up very well. Is there any probability that later in the season this seed will germinate, or has it rotted in the ground? Water stands within three feet of the surface and has considerable alkali. What can I plant on this land and get a crop? It is our intention to sow it to alfalfa next fall. The land adjoining, although higher, has a good stand of alfalfa now.

You are right about covering the alfalfa seed too deeply. It is not likely to appear. Your chance of getting a durable stand of alfalfa on such shallow soil over alkali water is not good, but you can hardly determine that without trying. Sometimes conditions are better than you think; sometimes worse. The plant itself is the best judge. On your lower land you could probably get a better stand of rye grass than anything else - sowing at the beginning of the rainy season. Of course, however, even that will depend upon how much alkali you have to deal with.

Alfalfa on Adobe.

Is adobe land good for alfalfa? Is it harder to start than in other soils or not? How much seed is required to sow an acre? Also state what time alfalfa should be sowed.

Alfalfa will thrive on an adobe soil if the moisture is kept right - especially guarding against too much water at a time. It is necessary to irrigate more frequently and apply only as much as can be absorbed by the soil before the hot sun comes on the field, for that scalds the plant badly. It is harder to get a good stand because of the cracking and hardening of the surface. Sow about 20 pounds to the acre just as soon as the soil comes into good condition - that is, moist and warm. February and March are usually the best months, according to the season in the interior valleys.

Alfalfa and Soil Depth.

Do you consider soil which is from 4 to 6 feet deep to hardpan of sufficient depth for alfalfa? Is there hardpan in the region of Lathrop in San Joaquin county, and can it be dissolved by irrigation, or can any good be accomplished by blowing holes at different places to allow the water to pass to lower levels? Are other crops affected by hardpan being so close to the surface?

You can grow alfalfa successfully on land which is from four to six feet deep if you irrigate rather more frequently and use less amounts of water each time, so that the plant shall be adequately supplied and yet not forced to carry its roots in standing water. The Eastern alfalfa grower is fortunate when he gets half the depth you mention, although it does seem rather shallow in California. Shallow lands are distributed over the valley quite widely. A deepening of the available soil is usually accomplished by dynamiting, especially so if the hardpan is underlaid by permanent strata. Alfalfa will penetrate some kinds and thicknesses of hardpan when it is kept moist, but not too wet, to encourage root growth.

Winter-growing green crops are less affected by shallow soil because they generally make their growth while the moisture is ample, if the season is good.

Curing Alfalfa with Artificial Heat.

It is current rumor that "out in California they are hauling alfalfa green and curing it by artificial heat," thus reducing loss through bad weather and producing a superior hay for feeding or milling purposes.

It is true that alfalfa is being cut green and dried by artificial heat, but this is only being done in preparation for grinding. No one thinks of doing it for the making of hay for storage or for feeding. This method is undertaken, not because the alfalfa hay does not dry quickly enough in the field, but because after drying in the field so many leaves are lost in hauling to the mill. We have no trouble sun-drying alfalfa for ordinary hay purposes; in fact, we have to be very careful that it does not get too dry.

Cheap Preparation of Land for Alfalfa.

I am about to put a piece of land into alfalfa, and want to use the most economical system of preparing the land for irrigation. My neighbors tell me that it will be necessary for me to have the land leveled; at a cost of $6 to $10 per acre. Now I am informed that in Alberta, and some places in California, they do not go to the expense of leveling land, but use a system of preparing land for irrigation at a cost of about 60 cents per acre.

Nothing except a highly educated gale of wind, with discriminating cutting and filling ability of a very high order, could do it for that price. The cheapest way to prepare land for irrigation is the contour check method, which is largely used, or the flooding in strips between levees at right angles to the supply ditch; but neither of these could be put in properly for that money, even if the land was naturally in such shape that a minimum amount of soil-shifting is necessary.

Where Alfalfa is Grown.

In what counties is alfalfa most successfully grown? By this I mean where three crops of hay may be had each growing season. Also, will corn grow good paying crops in same sections?

Alfalfa is grown all through the valleys and foothills of interior California; also to a certain extent in coast valleys. On suitable lands, three crops can sometimes be secured without irrigation, while twice or three times as many cuttings are secured on irrigated lands where the frost-free season is particularly long. According to the last census, we are growing alfalfa on 19,104 farms with a total acreage of 484,098. The total value of the product is over $13,000,000. Corn is widely grown, but is small as compared with alfalfa. It is grown in alfalfa districts and in coast valleys where there is not much done with alfalfa.

Sowing Alfalfa.

What is the proper time to sow alfalfa? Some advocate fall and others spring sowing. What seasons are given for each sowing?

We shall undoubtedly soon get to sowing alfalfa all the year round except in the short season of sharp frosts and cold wet ground in November, December and January. If you can get a good start in September and October, all right; if not, wait until February and March, according to the season. Where it is never very cold or wet, sow whenever moisture is right. There never can be any rule about it, for localities will differ.

Foxtail and Alfalfa.

Will foxtail choke out and exterminate alfalfa? Some fields look as though the foxtail had crowded the alfalfa out, but I hold that the alfalfa died from some other cause and the foxtail merely took its place.

Foxtail will not choke out alfalfa, providing, soil and moisture conditions are right for the latter, and a good stand of plant has been secured. If anything is wrong with the alfalfa, the foxtail will be on the alert to take advantage of it. You will always have foxtail with you, and considerable quantities of it, perhaps, in the first cutting, because foxtail will grow at a lower temperature than alfalfa, and, therefore, will keep very busy during the rainy season, while the alfalfa is more or less dormant, but as the heat increases, if the soil is good and moisture ample, the alfalfa will put the foxtail out of sight until the following winter invites it to make another aggressive growth. Therefore, we answer that alfalfa does not die from foxtail, but from some condition unfavorable to the alfalfa, which must be sought in the soil, or in the moisture supply, or traced back to bad seed, and a poor stand at the beginning.

Which Alfalfa is Best?

I have in Stanislaus county ten acres of Arabian alfalfa, which was sown the first week in April this year. It was clipped in July and irrigated. It is now about 14 inches high, but looks sickly, turns white at the tips, and some dies down. There are several places here with the Arabian alfalfa on them and with the same trouble, while the ordinary variety is looking fine by the side of it.

Arabian alfalfa usually makes a good show at first and begins to run out afterward. It does not seem to be so long-lived and satisfactory as the common variety. With this prospect ahead of you, according to present experience, it would seem to be desirable to plow the crop in and seed again with the common variety, or with the Turkestan, which is proving the most satisfactory of the recently introduced varieties.

Fall Sowing of Alfalfa.

We have summer-fallowed land which we know will grow good alfalfa, and as we have just had four inches of rainfall upon it, we were wondering if we could not plow the twenty acres and get a stand upon it in time to stand the cold weather this winter. Do you think this is practicable?

If four inches of rain on summer fallow connects well with the lower moisture which a good summer fallow ought to conserve in the soil, such sowing is rational; but if the summer fallowing was not done well, that is, if it was rough plowing without enough harrowing, as is too often the case, the four inches of rain might not be safe because of the dry ground beneath waiting to seize the moisture and so dry the surface that sprouting alfalfa plants would perish between dry soil below and dry wind above. Fall sowing will give enough growth to resist frost killing in many places in the valley if the moisture in the soil is enough to carry the plant as well as start it, or if showers come frequently - otherwise it is dangerous, not from frost but from drouth.

Alfalfa Hay and Soil Fertility.

We are feeding all our hay to dairy cows, returning the manure to the soil. At present prices of hay, my neighbors who sell theirs, seem to be as well off, with considerable less work; but how about the future? Can this soil be cropped indefinitely and the crops sold, without returning anything to the land?

It is a mistake to think that you can sell alfalfa hay indefinitely without reducing the soil. It may gain in nitrogen by the wastes of the plant, but it will lose in other constituents unless reinforced by fertilization. No single act can make for the maintenance of the soil as the growing and feeding of crops and return of manure does.

Dry-Land Alfalfa.

I am in a country of strictly dry farming. I have a wash or gulch on my place and would like to know if I could, with success, plant it to alfalfa without irrigation; soil is sandy loam, no evidences of springy moisture at all. What kind should I try?

Alfalfa will endure much drouth. What it will do in a particular place can only be told by trying. Sow Turkestan alfalfa. If the rains come early so as to wet the land down in September and October, sow the seed then. The endurance of the plant will depend much upon its having a chance to root deeply before the drouth comes on.

Inoculating Alfalfa.

206

Is it profitable to inoculate alfalfa seed before planting to increase its yield? Can it be done by leaching soil from old alfalfa ground, providing it has been plowed up and allowed to stand for a year? Are commercial inoculants a safe thing to inoculate with?

Apparently alfalfa does not need inoculation in this State. Probably not one acre in ten thousand now profitably growing alfalfa has ever had artificial introduction of germs. You can make germ-tea, if you wish, of the soil you describe; one year's exposure would not destroy the germs. It is safe enough to use commercial cultures. You will have to decide for yourself whether it is worth while.

Irrigating Alfalfa.

I am making parallel ridges for alfalfa, sending a full head of water down to the end of the field between each ridge. Should I calculate the lands to be mowed one at a time in even swaths? The mower being 5-foot cut, would you count on cutting a 4 1/2 or 5-foot swath? This soil is sandy, water percolating rapidly. The fall is 8 feet to the mile. How wide, then, would you advise making the ridges to suit the mower, and to flood economically, using from 2 to 4 cubic feet per second? The length of the lands is across 40 acres.

Growing alfalfa in long parallel checks, to be flooded between the levees, is the way in which much alfalfa is being put in at the present time where the land has such a slope as you indicate. It is calculated, however, to seed the levees as well as the check bottoms, and to run the mowers across the levees, thus leaving no waste land and mowing across the whole field and not between the levees as you propose. For that purpose these levees are made low, not over a foot in height, calculating that they will settle to about six or eight inches, which is sufficient to hold the water and direct its flow gently down the slope. There is, however, a limit to the distance over which water can be evenly distributed in this way, the difference being dependent upon the character of the soil, slope, etc. A length of nine hundred feet is sometimes found too great for an even distribution, and, for this reason, supply ditches at shorter intervals are introduced.

Unirrigated Alfalfa.

In what part of the State does alfalfa grow best without irrigation?

Obviously the parts which have the greatest rainfall in connection with retentive soil and plenty of summer heat. Alfalfa grows best without irrigation on "sub-irrigated" land where the ground water is sufficiently deep to allow a deep rooting of the plant in free soil and yet not too far down to be readily reached by the deep-running roots. Good results can be obtained with anywhere from four to ten or twelve feet of soil above water. On shallower soils the plant is apt to be short-lived through root troubles. Unirrigated alfalfa is also reduced by the incursions of gophers which flooding at least once a year will destroy.

Alfalfa and Overflow.

How long can alfalfa stand water without being drowned out? I have a piece of alfalfa on which the water will stand for considerable time in the winter time.

Alfalfa while dormant will endure submergence for several weeks. We do not know exactly how long, but evidently for a considerable period, providing temperatures are too low to invite growth. On the other hand, growing alfalfa is quickly and seriously injured by overflow.

No Nurse-Crop for Alfalfa.

Is it advisable to use oats with alfalfa seeds in seeding for alfalfa? Some growers of alfalfa here advise it strongly, others advise against it.

The general experience in California is decidedly against using oats, barley, or any other nursecrop with alfalfa. Get the land in the best possible condition and let the alfalfa have the full benefit of it. The ripening of the grain crop will do the young alfalfa plants more harm by robbing them of moisture than any protection which the taller plant can afford.

Reseeding Alfalfa.

This spring I planted alfalfa and only got about half a stand on some of the land. I want to reseed this fall and I thought of putting more seed on the ground and then disc it in. Or would you advise replanting the land? What do you think of putting manure on young alfalfa? Do you think there is any danger of burning it out?

Stir it up with a spring tooth harrow or disc it lightly to make a nice seed bed and then sow your seed as if you were planting alfalfa for the first time. This will give you a good seed bed and will not hurt the alfalfa already growing. Prepare the surface first and then sow, rather than disking in the seed. The manure in moderate application would not burn out the young alfalfa if properly applied after the rains begin.

Taking the Bloat Out of Alfalfa.

Will Italian rye grass and red top clover be a success under irrigation as cow pasture in this county, either separately or mixed with alfalfa? To sow in bare spots in the alfalfa, would the rye grass prevent bloat?

Italian rye grass and red clover will make good pasturage under irrigation and will make a fight with the alfalfa to the best of their ability. The admixture of rye grass will reduce the danger from bloating. Red clover will not have that effect, because red clover is a pretty good bloater on its own account. This seems to be the function of all the clovers according to the rankness of their growth at the time that they are grazed.

The Time to Cut Alfalfa.

What is the best period to cut alfalfa hay for cow feed and the best method for curing?

The best time to cut alfalfa is just when new shoots are starting out at the crown. This will give the greatest yield of hay during a season, and the hay will be much more palatable than if the alfalfa is permitted to get well into the blossoming period. The leaves, which are the best part of the hay, also remain on better than if the stems are older. If a person does not care to take the trouble to find out whether the new shoots are coming out or not, he can approximate the time to cut fairly well by waiting until a blossom here and there appears, cutting immediately. It would be difficult to tell on paper exactly when alfalfa was properly cured, as that is a matter of individual judgment. It is usual to cut in the morning and rake into windrows in the afternoon. With the usual weather in interior California that stage of the curing is completed by that time. The next day it can be gathered into cocks and gotten ready to move. That is about all the curing that is done. The size of the windrows depends upon the amount of hay, as thick hay should be put up in small windrows to give plenty of circulation of air. It is considered better also to build the cocks on raked land, otherwise the hay lying flat at the bottom will not cure properly and cannot be gathered up clean.

Which Crop of Alfalfa for Seed?

Which cutting of alfalfa should be left for seed bearing?

Which cutting is best for seed depends, of course, on the way the plant grows in your locality. Where it starts early and gives many cuttings in a season with irrigation a later growth should be chosen for seed than with a short season where fewer cuttings can be had. The second cutting is best in many places, but O. E. Lambert of Modesto after threshing about 30 lots in one year tells us that some growers had left second, some third and some fourth cuttings for seed. He found the second cutting very poor both in yield and grade, much of it not being well filled and the seed blighted, as the growth of hay was too heavy. The seed on third cutting was good both in grade and yield. Much of the seed on fourth cutting was not matured. For good results the stand should be thin. Our drier, heavier lands give the best results, sub-irrigated lands not seeding. All irrigation should stop with the previous cutting for hay.

Siloing First Crop Alfalfa.

How about putting first cutting of alfalfa and foxtail into the silo? Do you think there is any danger of fire in a wooden silo, and do you add salt and water when filling, and how long after it is cut would you advise putting it into the silo?

Put it through the silo cutter as soon as you can get it from the field. Do not let it cure at all, and be sure to cut and pack well. If at all dry, use water at the time of filling, and some salt then also, if you desire. There is no danger of firing if you put it in with good moisture, and by short cutting and hard packing you exclude the air. If you do not do this you will get a silo full of manure, and possibly have a fire while it is rotting.

Soil for Alfalfa.

What kind of soil is best for alfalfa on a dairy ranch?

An ideal soil for alfalfa is a deep well drained soil into which the roots can run deeply without danger of encountering standing water or alkali. Still we are finding that alfalfa is very successful on soils which are not strictly ideal, providing the moisture is supplied in such a way that the soil shall not be waterlogged nor the water be allowed to remain upon the surface during the hot weather, because this kills the plant.

Handling Young Alfalfa.

I have alfalfa that is doing very well for the first year. My soil is sandy loam with light traces of white alkali, although it does not seem to be detrimental to the growth thus far. I am in the dairy business and will have by winter enough manure to top-dress the field. Would it be good policy to use the manure, or would it be more satisfactory to top-dress with gypsum? Would it injure alfalfa to pasture lightly after the last cutting?

Presumably your soil contains enough lime, and therefore the application of gypsum at this time of the year would not be necessary. It may be desirable to top-dress with gypsum near the end of the rainy season to stimulate the growth of the plant. Gypsum, however, has no effect upon white alkali. So far as alkali goes, gypsum merely changes black alkali into white, thus making it less corrosive.

There would be no objection to pasturing lightly this fall. Be careful, however, to keep off the stock while the land is wet and not to overstock so as to injure root crowns by tramping. The manure can be used as a top dressing during the rainy season, unless you think it better to save it for the growth of other crops. Alfalfa is so deep rooting where conditions are favorable that it does not require fertilization usually on land which has been used for a long time for grain or other shallow-rooting plants.

Alfalfa Sowing with Gypsum.

I intend sowing alfalfa this fall on land that has some very compact hard spots. I aim to doctor these spots with gypsum at the rate of about 1000 pounds per acre and cultivate the gypsum in thoroughly two or three weeks before sowing the alfalfa seed. Would this be all right? Is there danger of injury to seed by coming in contact with gypsum?

Gypsum will not hurt the alfalfa seed. It is not corrosive like an alkali. Whether it will have time enough to ameliorate the soil in the spots in the period you mention depends upon there being moisture enough present at the time.

Red Clover for Shallow Land.

What can you say of red clover on shallow soils in the Sacramento valley under irrigation? How many crops, etc.?

Red clover is fine under the conditions you describe. We could never understand why people do not grow more of it on shallow land over hardpan which is free from alkali and not irrigated too much at a time. It is good on shallow land over water, where alfalfa roots decay, etc. Though we have no exact figures, we should expect to get about two-thirds as much weight from it as from an equally good stand of alfalfa.

Clovers for High Ground-Water.

Where, in California, is alfalfa being raised successfully above a water-table of, say, 4 feet or less, and are any unusual means used to accomplish this?

Over a high water-table, the alfalfa plant will be shorter lived according to the shallowness of soil above water. One could get very good results at from 4 to 6 feet, whereas at 2 or 3 feet the stand of alfalfa would soon become scant through decay of its fleshy root. Where the water comes very near the surface, a more shallow and fibrous rooting plant, like the Eastern red clover, should be substituted for alfalfa in California. It is a very vigorous grower and will yield a number of crops in succession although the water might be very near the surface, as in the case of the reclaimed islands in the Stockton and Sacramento regions and in shallow irrigated soils over bedrock in the foothills or over hardpan on the valley plains. In this statement, freedom from alkali is presumed.

Vetches in San joaquin.

In Michigan I was familiar with the use of the sand vetch as a forage plant, for hay, for green manure, and as a nitrogen producer.

214

In western Michigan, on the loose sandy soil, I sowed in September or October 20 pounds per acre for a seed crop and 40 pounds per acre for pasture, hay, or green manure. Can I expect good results in Fresno and Tulare counties without irrigation? Will fall seeding the same as wheat produce a seed crop? Will sand vetch grow on soil having one-half of one per cent alkali?

Most of the vetches grow well in the California valleys during the rainy season; the common vetch, Vicia sativa, and the hairy vetch, Vicia hirsuta, are giving best results. The proper time to plant is at the beginning of the rainy season. They will stand some alkali, especially during the rainy season, when it is likely to be distributed by the downward movement of water, but it is very easy to find land which has too much alkali for them. These plants seed well in some parts of the valley, but a local trial must be made to give you definite information.

Growing Vetch for Hay.

How many pounds of vetch seed should be sown to the acre? How many tons per acre in the crop? As I desire to change my crop, having to some extent exhausted the soil with oats, how advisable will it be to sow wheat with the vetch to give it something to climb on? If so, and wheat is not desirable under the circumstances, what? In using vetch for horse fodder, how much barley should be fed with it per day for a driving horse? For a draught horse? Is vetch sown and harvested at about the same time as other crops?

Except in very frosty places, vetch can be sown after the rain begins at about 40 to 60 pounds of seed to the acre. The yield will depend upon the land and on the moisture supply, and cannot be prophesied. One grower reports three tons of hay per acre near Napa. If the land usually yields a good hay crop, it should yield a greater weight of vetch. In mowing for hay purposes it is desirable to raise the vetch off the ground to facilitate the action of the mower. Oats would be better than wheat, because rather quicker in winter growth. If the vetch is to be fed green, rye is a good grain, but not good for hay purposes because of the hardness of the stem. There is no particular difference in the plant-food requirements of the different grains, so that there is nothing gained in that way in the choice of wheat. In feeding a combined vetch and barley hay, the ration is balanced; the feeding of grain would not be necessary, except in case of hard work under the same conditions grain is usually fed to horses and in about the same amounts. Vetch requires a longer season than ordinary oat or barley hay crop to make a larger growth, consequently an early sowing is desirable.

Cover Crop in Hop Yard.

Will you please give information concerning cow peas or the most suitable crop to sow in a hop field for winter growth, to be plowed under as a fertilizer in the spring? Also, would it injure the vines to be cut down before they die, so as to sow the mulch crop soon as possible after the hops are gathered?

Cow peas would not do for the use which you propose, because they would be speedily killed by frost on low lands, usually chosen for hops, and would give you no growth during the frosty season. Probably there is nothing better than burr clover for such a winter growth. Hop vines should be allowed to grow as long as they maintain the thrifty green color, because the growth of the leaves strengthens the root. But when they begin to become weakened and yellow they can be removed without injury. It is not necessary to wait for them to become fully dead.

Growing Cowpeas.

What is the best variety of cow peas for a forage crap? I want a variety which with irrigation will come up after it has been cut, so as to keep growing and not be like some which I tried last year. They grew up like ordinary garden peas and were just a waste of ground.

Possibly you did not get cowpeas; they do not look like garden peas at all: they look more like running beans, which they are. The crop is not counted satisfactory except on low, moist land, for on uplands, even with irrigation, it does not seem to behave right. We do not know that a second growth can be expected, for in the Southern States it is grown as a single crop, and resowing is done if a succession is desired, the point being made at the South that the plant is adapted to this method of culture because it grows so rapidly that it can be twice sown and harvested during the frost-free period.

Cowpeas in the San Joaquin.

How late in the season will it be profitable to plant cowpeas? What is the best manner of planting? Are there several varieties? If so, which one is best adapted to plant after oats? The land can be irrigated until about August 10. Will it be advisable to plow up a poor stand of alfalfa about July 1 and plant to cow peas?

You can plant cowpeas all summer on land which is moist enough by natural moisture or irrigation to promote growth. What you will get by late planting depends upon moisture and absence of an early fall frost. If your alfalfa stand is bad enough to need resowing anyway, you may get a good catch crop of cowpeas by doing as you propose. If, however, you plow under much coarse stuff in putting in the peas the growth may be irregular. It can, of course, be improved by free irrigation. On clear land moderately retentive much more is being done in summer growth of cowpeas without irrigation than expected. There are several good varieties. One of these is the Whippoorwill. Cowpeas can be sown in furrows three feet apart and cultivated, using about 40 pounds of seed to the acre, or they may be broadcasted, which takes about twice as much seed.

Cowpeas and Canadian Peas.

Would Canadian field peas and cow peas be valuable as a forage crop for cows and hogs; also as fertilizer? Please tell us also when to plant, how to plant, etc.

These plants are of high forage value as cow feed; also as a soil restorative when the whole crop is plowed under green or when the roots and manure from feeding add to the soil. But for either purpose the result depends upon how much growth you can get, and that should be told by local trial before any great outlay is undertaken. Canadian peas are hardy against frost and can be broadcasted and covered with shallow plowing as soon as the land is moist enough from fall rains - except in very frosty parts of the State. They can also be sown in drills to advantage. Cow peas are beans, and cannot be planted until frost danger is over in the spring. They are only available for summer feeding, and whether they will be worth while or not depends upon how much moisture can be held in the soil for summer growth. They should be sown in drills and cultivation continued for moisture conservation until the plants cover the ground too much to get the cultivator through.

Canadian or Niles Peas.

I send a sample of peas which I bought for Canada field peas, and they were so labeled. I would like to know what they are.

The peas are, apparently, one kind of Canada peas. There is some variation in Canada peas, but these are peas of that class. Some of the Canada pea are hardly distinguishable from the so-called Niles pea of California growth, and it does not matter much, anyway, for one is about as good as the other.

Sunflowers and Soy Beans.

I would like information concerning cultivation, method of feeding and food value of soy beans. Also sunflowers.

Soy beans are grown like other beans, in rows which, for convenience in field culture, should be about 2 1/2 feet apart and cultivated up to blooming time at least. They should be sown after frost danger is over and the weather is settled warm, for they enjoy heat. For feeding they can be made into hay before maturity, or the beans can be matured and prepared for feeding by grinding. As with other beans, small amounts should be used in connection with other feeds. They are a rich food and somewhat heavy on the digestion. The same is true of sunflowers, except that the seed is richer in oil than in protein, as beans are. Sunflowers in field culture are planted and cultivated like beans. The seed is flailed out of the heads after they lie for a time to dry.

Jersey Kale.

Please inform me how to plant Jersey or cow kale.

Jersey kale can be planted by thin scattering of seeds in rows 2 1/2 feet apart so as to admit of cultivation, or the plants can be grown just as cabbage plants are and set out 2 1/2 or 3 feet apart, the squares to admit of cultivation both ways. The plant needs a good deal more space than an ordinary cabbage, for it makes a tall free growth, and space must be had for the growth of the plant and for going into the patch for stripping off leaves and cultivation. The plant can be started in the rainy season whenever the land comes into good condition. It is a winter grower in California valleys.

Rape and Milo.

Would rape be a good pasture crop sown broadcast? If so, at what time should it be planted? Will Milo maize grow profitable in Sonoma county?

Rape can be sown as soon as the land gets moist enough from early rains to start the seed and hold the growth. It is a wintergrowing plant in this State. We believe, however, you will get better results with common vetch, which is also a winter grower and more nutritious. If you desire one of the cabbage family, kale will probably serve you better than rape. Milo is one of the sorghums and will only grow during the frostless period, like Kafir, Egyptian corn and other sorghums. It will do well with you, but probably make less growth than in the interior valleys.

Sweet Clover Not an Alfalfa.

I send you a sample of alfalfa which grows very vigorous here on my place spontaneously and would like you to give me all the information about it you will, as a feed for cows and hogs. The stock seem to eat it well.

The plant is not alfalfa at all. It is white sweet clover (melilotus alba), and it is usually considered a great pest in alfalfa fields, because although it grows vigorously as you describe, it is not generally accepted by stock, unless once in awhile some one considers it a good thing, perhaps because he keeps stock hungry enough to enjoy it in spite of its rank taste and smell, but, usually when they can get alfalfa they will not pay much attention to this plant. It is good for bee pasturage, however, and is grown to some extent for that purpose. You probably had the seed of it in your alfalfa seed. It is a biennial and not a perennial like alfalfa. It will disappear if you can keep it from going to seed.

Sweet Clover as a Cover Crop.

How about melilotus as a cover crop? Last year in certain sections it proved very successful, while in others it did not give satisfaction.

Melilotus, by virtue of its hardiness in growing at low temperatures, its depth of root penetration, the availability of the seed, the smallness of the seed so that the weight required for the acre is not large, is to be favored for a cover crop. The objections are two: The fact that it does not seem to grow well under some conditions; second, that when a growth is made it is coarse and rangey, and the amount of green stuff to the acre is much less than its appearance would indicate. We know of cases where what seemed to be a good stand of melilotus yielded only about ten tons of green stuff to the acre, and what appeared to be a less growth of vetches or peas yielded from fifteen to twenty tons to the acre. And yet we believe that in some places it will be found extremely desirable for a cover crop in harmony with what was reported some time ago as the result of experiments by the Arizona Experiment Station.

Spineless Cactus.

There seems to be two distinct kinds of cactus: One for forage, the other for fruit. It is claimed by some people that the spineless cactus is more valuable as a forage plant than alfalfa. What is your opinion?

There are many varieties of smooth cacti. Some of them bear higher quality fruit than others, and some are freer growers and bear a greater amount of leaf substance for forage purposes; therefore, varieties are being developed which are superior for fruit or for forage, as the case may be. Spineless cactus is in no way comparable with alfalfa, either in nutritive content or in value of crop, providing you have land and water which will produce a good product of alfalfa. Cactus is for lands which are in an entirely different class and which are not capable of alfalfa production.

Probably Not Broom-Corn.

I have a side-hill ranch on which I would very much like to raise broom corn. The soil produces good grapes, fruit, corn, oats, peas, etc., and I wish to know if there are possibilities of broom-straw.

All the broom-corn which has been successfully produced in California has been produced on moist, riverside land. The plant is a sorghum - consequently subject to frost injury, and can only be grown during the frostless season as Indian corn is. This makes it impossible to get the advantage of rainfall on winter upland and necessitates the use of lowlands, which carry moisture enough to secure a free growth of the brush, for poor broom-corn is worthless practically, being too low priced to be profitable for brooms and too fibrous to be of value for feeding purposes. Even in a place where the plant grows well its product is worthless unless properly treated, and that requires full knowledge and a good deal of work.

The Outlook for Broom Corn.

Broom corn is way up in price, but that is an indication that everyone who has ever grown broom corn is likely to plant it this year. What is the outlook in California?

Nothing but a local experiment will determine whether you can get a satisfactory brush under the conditions prevailing in your vicinity. Undoubtedly, the high price of broom corn will stimulate production, but under quite sharp limitations in California, because a good, satisfactory brush cannot be grown on dry plains, although a good product is made in the river bottoms not far away. But there are so few people in California who understand how to handle broom corn to produce a good commercial article, and there are such rigid requirements in the size, quality, etc., that those who break into the business without proper knowledge cannot command even profitable prices. Therefore, if your enterprise is conducted with a full knowledge and with proper local conditions it would not encounter such a local disadvantage in the great increase of the product as one might think at first.

Smutty Sorghum.

The various plantings of Egyptian corn on the ranch have turned smutty, very much after the manner of wheat and barley. Is there any unusual reason for this, or could irrigation have caused it, and what is the best method of preventing it?

Sorghum is affected by a smut similar to that of other grains. It is due to the introduction of the germ of the disease which comes with the use of smutty seed. Possibly the growth of the smut may have been promoted by moisture arising from soil rendered very wet by irrigation, and for this plant free irrigation should not be used, because it will do more with less water than any other plant we are growing, and is likely to be more thrifty in a drier atmosphere. Get seed for next year from an absolutely clean field; get as much growth as you can without irrigation, and then use water in moderate quantities as may be necessary, followed by a cultivation for the drying of the surface.

Late-sown Sorghum.

How late can Egyptian corn be planted on good sediment soil capable of growing 40 to 50 socks of barley per acre in good years with ordinary rain? The field was cut this year for hay on account of rank growth of wild oats, after irrigating; land is still moist. Can I put in Egyptian corn with on assurance of crop, or is it too late? How much seed should be planted to the acre, also should seed be drilled in or broad-casted?

There is no difficulty in getting a start of Egyptian corn during the dry season providing the soil contains moisture enough to germinate the seed. Afterward the growth will be more or less according to the moisture present and will be available for forage purposes. Whether a seed crop can be had by late sowing depends upon the frost occurrence in the particular locality, for it only takes a light frost to destroy the plant. To get the best results, particularly with late sowing, the seeds should be drilled in rows far enough apart for horse cultivation; about forty pounds of seed to the acre. What you get in this way will depend upon the amount of moisture in the soil and the duration of the frost-freedom.

Kaffir and Egyptian Corn.

Does Kaffir corn yield as well here as Egyptian corn? The fodder is good feed and the heads stand erect and at a more even height from the ground, which makes three advantages over Egyptian. Irrigation in either case is the some.

The reasons you mention have no doubt had much to do with the present popularity of an upright plant like Kafir over a gooseneck like the old dhoura or Egyptian, which was the type first introduced in California. For years there has been more gooseneck sorghum in the Sacramento valley than in any other part of the State. It may have superior local adaptions or the people may be more conservative. The way to determine which is better is to try it out, and, unless the Egyptian does better in grain and forage than the upright growers, take to the grain which holds its head up.

Sorghums for Seed.

Which sorghum is the most profitable to plant for the seed only White
Egyptian, Brawn Egyptian or Yellow Mila?

Which sorghum is best is apparently a local question and governed by local conditions to a certain extent. Egyptian corn (with the goose-neck stem) has held more popularity in your part of the Sacramento than elsewhere, while Kaffir corn (holding its head upright, as do many other sorghums) has been for years very popular in the San Joaquin. In the Imperial valley Dwarf Milo is chiefly grown for a seed crop shattering and bird invasion are very important. G. W. Dairs of the San Joaquin valley, says there is a very great difference in the different varieties regarding waste from the blackbird. The ordinary white Egyptian corn is very easily shelled, and the birds waste many times more of the grain than they eat, after it has become thoroughly ripe. The Milo maize, or red Egyptian corn, does not shell nearly so easily as the white corn, and the grain is considerably harder and less attractive to the blackbirds. In fact, blackbirds will not work in a field of this variety of corn if there is any white corn in the vicinity to be had. The dwarf Milo maize yields much more crop than the white Egyptian corn, or any other variety. Blackbirds do not damage the white Kaffir corn to the extent they do the ordinary white Egyptian corn.

Sorghum Planting.

What is the best time to sow Egyptian corn; also how much per acre to sow?

All the sorghums, of which Egyptian corn is one, must be sown after frost danger is over - the time widely known as suitable for Indian corn, squashes and other tender plants. Sow thinly in shallow furrows or "marks," 3 1/2 or 4 feet apart and cultivate as long as you can easily get through the rows with a horse. About 8 pounds of seed is used per acre. If grown for green fodder, sow more thickly and make the rows closer, say 2 1/2 feet apart.

Buckwheat Growing.

Two or three farmers in this locality desire to plant buckwheat. Not having done so heretofore they are in doubt as to the soil and other conditions that go to make a successful crop.

The growing of buckwheat in California is an exceedingly small affair. The local market is very limited, as most California hot cakes are made of wheat flour. There is no chance for outward shipment, and the crop itself, being capable of growing only during the frostless season, has to be planted on moist lands where there is not only abundant summer moisture but an air somewhat humid. Irrigated uplands, even in the frostless season, are hardly suitable for the common buckwheat, although they may give the growth of Japanese buckwheat for beekeepers who use dark honey for bee feeding. The Japanese buckwheat is well suited for this because it keeps blooming and produces a scattered crop of seed, but this characteristic makes it less suitable for a grain crop, and it has therefore never become very popular in this State. We consider buckwheat as not worthy of much consideration by California farmers.

Variation in Russian Sunflowers.

In an acre of mammoth Russian sunflowers there seems to be three varieties, some of the plants bear but one large flower; others bear a flower at the top with many other smaller ones circling it, while others have long stalks just above the leaf stems from the ground level all the way up to the largest flower, which appears at the very top. Are all these varieties true mammoth Russian sunflowers? What explanation is there for these variations? Will the seed from the variety carrying but one natural head produce seed that will reproduce true to the parent?

Your sunflowers are probably only playing the pranks their grandfathers enjoyed. If seed is gathered indiscriminately from all the heads which appear in the crop, succeeding generations will keep reverting until they return to the wild type, or something near it. If there is a clear idea of what is the best type (one great head or several heads, placed in a certain way) and seed is continually taken from such plants only for planting, more and more plants will be of this kind until the type becomes fixed and reversions will only rarely appear. No seed should be kept for planting without selecting it from what you consider the best type of plant; no field should be grown for commercial seed without rogue-ing out the plants which show reversions or bad variations. If you find sunflowers profitable as a crop in your locality, rigid selection of seed should be practiced by all growers, after careful comparison of views and a decision as to the best characters to select for.

Sacaline.

My attention has been brought to a plant called Sacaline by an Eastern plant dealer. He states that this plant will grow in any kind of soil and needs practically no water.

The plant Sacaline (Polygonum saghalience) was introduced to California as a dry-land forage plant about 1893, and has never demonstrated any particular forage value. It is a browsing shrub, making woody stem, and cattle will eat it readily when not provided with better food. It has possible value on waste land, but probably is in no sense superior to the native shrubs of California which serve that purpose. It is a handsome ornamental plant for gardens or parks.

Mossy Lawns.

What will destroy patches of moss which are spreading over our lawns and apparently destroying the grass?

More sunlight would have a tendency to discourage the growth of moss on a lawn. If this is not feasible, irrigation less frequently but a more thorough soaking each time will give the surface a better chance to dry off, and moss will not grow on a dry surface. The frequent spraying of a lawn with just enough water to keep the surface moist and not enough water to penetrate deeply will tend to the growing of moss and to less vigor in the growth of the grass, A good soaking of the soil once a week is better than daily sprinkling, but, of course, very much more water must be used when you only sprinkle at long intervals. The drying of the surface may be assisted by sprinkling with air-slaked lime and this will discourage the growth of moss, but of course lime must not be used in excess or it will also injure the grass.

Scattering Grass Seeds.

We live on the west side of Sonoma valley, and want to seed some of our fields with a good wild grass. We want to carry bags of it
in our pockets to scatter when we ride. Timothy we should like, but
this is not its habitat, is it? Can you suggest a grass or grasses that
would do well here?

There are really wild grasses worthy of multiplication, but no one
makes a business of collecting the seed for sale, so that such seeds
are not available for such purpose as you describe. Of the introduced grasses, those which are most likely to catch from early scattered seed are Australian and Italian rye grasses, orchard grass,
wild oat grass and red top. You can get seed of all these from dealers in any quantity which you desire at from 15 to 30 cents a
pound, according to the variety, and make a mixture of equal parts
of each grass, which you can carry and scatter as you propose. Some
of them will catch somewhere, particularly in spots where the shade
modifies the summer heat and where seepage moisture reduces soil
drought. You are right about timothy; it is good farther up the coast
and in the mountain valleys, but not in your district.

Poultry Forage.

I have light sandy loam on which I desire to grow forage for chickens.
It lies too high for irrigation.

You could probably grow alfalfa to advantage if the soil still deep
and loose, getting less, of course, than by irrigation, but still an
amount that would be very helpful in your chicken business.
Otherwise, as the land lies higher and perhaps out of sharp frosts,
you could grow winter crops of vetches and peas and thus improve

the land while furnishing you additional poultry pasture. The latter purpose could also be served by growing beets, cabbage or other hardy vegetables during the rainy season. This is prescribed because of the apprehension that the soil may not contain moisture enough for summer cropping without irrigation.

No Grain Elevators in California.

Is California wheat shipped in bulk or in bags at the present time?

There are no elevators in this State, owing to the fact that hitherto grain cargoes have been acceptable to ship only as sacked grain, because of claimed danger of shifting cargo and disaster during the long voyage around the Horn. A novel by Frank Norris, entitled the "Octopus," describes a man being killed by smothering in a grain elevator at Port Costa, but there never was an elevator at that point, and consequently there never was a man killed by getting under the spout thereof. Answering specifically your question, California grain is shipped in bags and not in bulk. It is handled in sacks from the separator to roadside or riverside storage, to the loading point into the ships and out of the ships on the other side - still in bags.

New Zealand Flax.

Give information about Phormiun tenax (New Zealand flax), which I see is imported to San Francisco in large quantities yearly for making cordage and binder twine, and is said also to be the best of bee pasture. Can I get the plants on the coast, and is California soil and climate adapted to the culture?

New Zealand flax grows admirably in the coast region of California. You will find it in nearly all the public parks and in private gardens, for it is a very ornamental perennial. Plants can be had in any quantity from the California nurserymen and florists. It produces plenty of leaves, but we should doubt whether it is floriferous enough for bee pasturage except where it occurs wild over a large acreage. You could get vastly more honey from other plants grown for that purpose.

No Home-made Beet Sugar.

Is there any simple process of making sugar from beets so that I could make my own sugar at home from my own beets while sugar is so very expensive to buy?

There is no simple way of making beet sugar. It can only be economically done in factories costing hundreds of thousands of dollars.

Don't Get Crazy About Special Crops.

I want information about flax as a crop. I have been having some land graded for alfalfa and I have had to wait so long I am now doubting the advisability of seeding it all under these conditions until fall, as hot weather will soon come. I want some good crop to plant in the checks and give two good irrigations. What would you think about rye for straw for horse collars? I do not wish to consider corn, as the stalks would be troublesome. Potatoes would necessitate disarranging the land too much and would require more attention than I am in shape to give just flow. Everybody grows wheat, barley and oats. I want something that I can get a special market for.

To succeed with flax, the seed ought to be sown in the fall, or early winter, in California, and the plant will make satisfactory growth under about the same conditions that suit barley or wheat. Spring sowing would not give you anything worth while except on moist bottom land. Rye is also a winter-growing grain. To grow rye straw for horse collars would be unprofitable unless you could find some local saddler who could use a little, and it is probable you could not get a summer growth of rye which would give good straw, even if you had a market for it. You could get a growth of stock beets, field squashes, or pumpkins for stock feeding. In fact, the latter would give you most satisfaction if you have stock to which they can be fed to advantage. Sorghum is our chief dry-season crop, but that makes stalks like corn and would, therefore, be open to the same objections. Has it never occurred to you that people grow the common crops, not because they are stupid, but because those are the things for which there is a constant demand and the best chance for profitable sale? Efforts to supply special markets are worth thinking of, but seldom worth making unless you know just who is going to buy the product and at what price.

California Insect Powder.

What part of the plant is used in making insect powder and how is it prepared? Is the plant a perennial? What soil suits it best?

The plant is Pyrethrum cinerariaefolium and has a white blossom resembling the common marguerite. The powder is made of the petals and the seed capsules or heads are thoroughly dried in the sun and ground with a run of stone such as was formerly used for making flour. The powder must be finely ground, and only good powder can be made in a mill suitably equipped for that purpose. The plant is a perennial, beginning to bloom the second year from seed. It will grow in any good soil with ordinary cultivation. Twenty-five years ago it was thought that a great California industry might be established on that basis, but there is at the present time but one establishment, which grows about all the material it can use on its own ranch in Merced county, on a fine, deep loam which the plant seems to enjoy.

Rotations for California.

I wish to work out a practical system of crop rotation suitable to the climate and conditions obtaining in southern California. Would you recommend different systems for grain lands and irrigated lands?

General schemes of rotation are hard to work out in California. They must be locally revised according to the local temperature conditions and the local market also. We should endeavor to find out what has been successfully grown on similar lands to those which you have in mind and arrange the rotation on that basis, from what we knew of the relation of the different plants to soil fertility, etc. You cannot make out a satisfactory local scheme for the seven counties in southern California, because of the widely different behavior of the separate plants in the different parts of the district. You can hardly work on the basis of soil character: moisture supply and temperatures are more determinative. Surely you should make a scheme for irrigated land different from that for dry land, and it could not only be a longer rotation, but many more plants would be available for its service.

Berseem?

Berseem has been introduced into this country from Egypt, and would like to know if it has been used in California, and if it has came up to expectations.

Berseem is an annual clover supposed to grow only during the summer time. It has been tried widely in California, but practically abandoned because it will not grow during the rainy season. It is in no way comparable to alfalfa, which is a deep rooted perennial plant, nor would it be comparable with burr clover as a winter grower on lands which have a moderate amount of water.

Heating and Fermentation.

Please explain why dampness will cause anything like hay, Egyptian corn or other like products to heat.

Heating is due to fermentation, which means the action upon the vegetable substance of germs which begin to grow and multiply after their kind whenever conditions favor them. The earlier stages of this action is called "sweating," and it is beneficial as in the case of hay, tobacco, dried fruits, etc., as is commonly recognized - resulting in what is known as curing - and it is the art of the handler of such products not to allow the action to go beyond what may be called the normal "sweating." If not checked by proper handling, which involves drying, cooling, etc., fermentation will continue, and other germs will find conditions suitable for them to take up their work of destruction, and this new action produces higher temperature still, and if not checked by cooling or drying or otherwise making the substance inhospitable to them, "heating" will result, and thence onward rapidly to decay, if they have everything their own way.

Moonshine Farming.

What influence, if any, has the moon on plant growth? Are there any reliable data of experiments available?

Very prolonged investigation by the Weather Bureau determined that no difference was found in planting in different phases of the moon. If we paid any attention to it, we should plant in the dark of the moon, so as to get the plants up so that they could use the little more light which the moon gives. It is, however, more important to have the soil right than the moon.

Part IV. Soils, Fertilizers and Irrigation

What is Intensive Cultivation?

From whom can I receive instruction or information regarding intensive cultivation?

Intensive cultivation has, so far as we know, not been made the subject of any treatise or publication. Intensive cultivation means the use of a maximum amount of labor, fertilizers and water for products of high market value. There is no better example of intensive cultivation in the world than is afforded by the practice of the best market gardeners and producers of small fruit. Next to them, on larger areas, would be the policies and methods of the fruit growers of California. Intensive culture, then, is not a particular method or system, but consists in doing the best thing for maximum production of any product which is valuable enough to spend the large outlay which is required. Just how this cultivation should be done depends upon the nature of the product and the conditions of soil and climate in whatever locality intensive cultivation may be undertaken.

Can a Man Farm?

Is it possible for a man with a few acres well cared for and carefully tilled to make a living and pay out on a purchase of land at $123 per acre? Could a good carpenter make wages and take care of a small tract for a year or so until well under way?

We consider $125 per acre for good land with a good water right a fair price. Financing a farming operation depends more upon the man than upon the good land. There are men who would, by intensive cultivation of salable stuff and right use of water, pay off the full value of the land from its produce in a couple of years. Others will never pay off. Of course, the nearer you can come to paying for the land at the beginning, and the more money you have for improvements, the more satisfactory your situation should be in every respect. There is a good chance for carpenter work in colony development, and considerable self-help could be secured in that way. You do not say whether you know anything about farming. Farming is a very complicated business and a basic knowledge derived from experience is a proper foundation to build upon in the light of the fuller application of scientific principles.

Soil Depth for Citrus Trees.

I have a top soil of rich loam containing small rocks and pebbles. Underneath it is washed gravel, rocks, boulders, yellow sand, etc. What is the limit as to thinness before trees will not grow, or thrive?

Orange trees are growing quite successfully on shallow soil overlying clay where the use of water and fertilizers was carefully adjusted so as to keep the trees supplied with just the right amount. Under such conditions a good growth may be expected so long as this treatment is maintained. There should be, however, not less than three feet of good soil to make the large expenditure necessary to establish an orange orchard permanently productive, and all the depth you can get beyond three feet is desirable. We question the desirability of planting orange trees even on a good soil overlying gravel, rocks or sand. Roots will penetrate such material only a short distance usually. It is almost impossible with such a leachy foundation to keep the surface soil properly moistened and enriched; You are apt to lose both water and fertilizer into the too rapid drainage.

Soils and Oranges.

I find this entire district underlaid with hardpan at various depths, from 1 to 6 feet down, and of various thicknesses. This hardpan is more or less porous and seeps up water to some extent, but is too hard for roots to penetrate. It is represented to me that if this hard pan is down from 4 to 5 feet it does not interfere with the growth of the orange tree or its producing. Is 4 or 5 feet of the loam enough?

Four or five feet of good soil over a hardpan, which was somewhat porous, is likely to be satisfactory for orange planting. There has been trouble from hardpan too near the surface and from the occurrence of a hardpan too rich in lime, which has resulted in yellow leaf and other manifestations of unthrift in the tree. Discussion of this subject is given on page 434 of the fifth edition of our book on "California Fruits," where we especially commend a good depth of "strong, free loam." This does not mean necessarily deep. The orange likes rather a heavier soil, while a deep sandy loam is preferred by some other fruits. If you keep the moisture supply regular and right and feed the plant with fertilizers, as may be required, the soil you mention is of sufficient depth - if it is otherwise satisfactory.

Oranges Over High Ground Water.

Does California experience show that citrus trees can be grown upon land successfully where the water-level is 6 feet from the surface; that is, where water is found at that level at all seasons and does not appear to rise higher during the rainy season?

We do not know of citrus trees in California with ground-water permanently at six feet below the surface. If the soil should be a free loam and the capillarity therefore somewhat reduced, orange trees would probably be permanently productive. If the soil were very heavy, capillary rise might be too energetic and saturate the soil for some distance above the water-level. In a free soil without this danger the roots could approach the water as they find it desirable and be permanently supplied. Orange trees are largely dependent upon a shallow root system, the chief roots generally occupying the first four feet below the surface. From this fact we conclude that deep rooting is not necessary to the orange, although unquestionably deep rooting and deep penetration for water are desirable as allowing the tree to draw upon a much greater soil mass and therefore be less dependent upon frequent irrigations and fertilizations.

Depth of Ground-Water.

Is there probable harm from water standing 12 feet from the surface in an orchard? Also probable age of trees before any effect of said water would be felt by them? The soil is almost entirely chocolate dry bog. - W. E. Wahtoke.

Water at twelve feet from the surface is desirable, and water at that point will be indefinitely desirable for the growing of fruit trees. Of course, conditions would change rapidly as standing water might approach more nearly to the surface, a condition which has to be carefully guarded against in irrigation. But it can come nearer than twelve feet without danger.

Summer Fallow and Summer Cropping.

I own some hill land which has been run down by continuous hay cropping. I am told that a portion must be summer-fallowed each year, but I wish to grow some summer crop on this fallow ground that will both enrich the soil and at the same time furnish good milk-producing feed for cows - thoroughly cultivating it between the rows. What crop would be best? I am told the common Kaffir or Egyptian corn are both soil enriching and milk producing.

If you grow a summer crop on the summer-fallowed upland, you lose the chief advantage of summer fallowing, which is the storing of moisture for the following year's crop. A cultivated crop would waste less moisture than a broadcast crop, surely, but on uplands without irrigation it would take out all the moisture available and not act in the line of a summer fallow.

Kaffir corn is not soil enriching. It has no such character. It probably depletes the soil just as much as an ordinary corn or hay crop. It is a good food to continue a milking period into the dry season, but you must be careful not to allow your cattle to get too much green sorghum, for it sometimes produces fatal results. We do not know anything which you can grow during the summer without irrigation which would contribute to the fertility of your land. If you had water and could grow clover or some legume during the summer season, the desired effect on the soil would be secured.

Soils and Crop Changes.

Peas and sweet peas do not grow well continuously in the same ground. I know this practically in my experience, but in no book have I ever found why they do not grow.

There are two very good reasons why some classes of plants cannot be well grown continuously in the same piece of ground. One is the depletion of available plant food, the other the formation of injurious compounds by the plants, or the gradual increase of fungoid, bacterial or animate pests in the soil, which finally become abundant enough to seriously hinder growth. Different plants take the plant foods, as nitrogen, lime, potash, phosphates, etc., in different proportion. More important, perhaps, is the fact that the root acids that extract these foods are of different types and strength.

Thus before many seasons it may happen that most of the plant food of one or more kinds may be nearly exhausted as far as that kind of plant is concerned that has grown there continually, while there would be plenty of easily available food for plants with a different kind of root system and different root acids, etc. This is one reason why rotation of crops is so good; it gives a combination of root acids and root systems to the soil during a term of years, and it also frees the soil from one certain kind of organism because it cannot survive the absence of the particular plants on which it thrives.

Summer-Fallow Before Fruit Planting.

I recently bought a ranch at Sheridan, Placer county, and was intending to put 10 acres to peaches and 50 acres to wheat or barley, but the residents tell me that the land must be summer-fallowed before I can do anything. The soil is a red loam and has not been plowed for six years.

Your local advisers are probably right as to the necessity for summer-fallowing in order to conserve moisture from a previous year's rainfall and to get the land otherwise into good condition. There might be such a generous rainfall that an excellent crop might come without summer-fallowing, and the results will depend upon the rainfall. If it should be small in amount, you might not recover your seed. By the same sign you might not get much growth on your fruit trees, but you could help them by constant cultivation and by using the water-wagon if the season should be very dry. Therefore, you are likely to do better with trees than with grain without summer-fallowing, although even for trees it is a decided advantage to have more moisture stored in the subsoil and the surface soil pulverized by more tillage.

Defects in Soil Moisture.

242

I have apricot trees that appear to be almost dead; all but a very few small green leaves are gone, and they look bad, still I think they might be saved if I only knew what to do.

Presumably your apricot tree is suffering from too much standing water during the dormant season, or from a lack of water during the dry season. The remedy would be to correct moisture conditions, either by underdrainage for winter excess or by irrigation for summer deficiency. When a tree gets into a position such as you describe, it should be cut back freely and irrigation supplied, if the soil is dry, in the house that the roots may be able to restore themselves and promote a new growth in the top.

Dry Plowing for Soil and Weed Growth.

Is there any scientific reason to support the belief that it is injurious to the soil to dry-plow it for seeding to grain this fall and winter? Will dry-plowing now cause a worse growth of filth after the rains than the customary fallowing in the spring? Should the stubble be burned, or plowed under!

The points against dry-plowing to which you allude may arise from two claims or beliefs: first, that turning up land to the sun has a tendency to "burn out the humus"; second, that dry-plowing may leave the land so rough and cloddy that a small rainfall is currently lost by evaporation and leaves less moisture available for a crop than if it is plowed in the usual way after the rains. The first claim is probably largely fanciful, so far as an upturning in the reduced sunshine of the autumn goes. Whatever there may be in it would occur in vastly increased degree in a properly worked summer-fallow, and even that is negligible, because of the greater advantage which the summer-fallow yields. There may be cases in which one will get less growth on dry-plowing than on winter plowing, if the land is rough and the rain scant, and yet dry-plowing before the rains is a foundation for moisture reception and retention - if the land is not only plowed, but is also harrowed or otherwise worked down out of its large cloddy condition. When that is done, dry-plowing may be a great help toward early sowing and large growth afterward. As for weeds, dry-plowing may help their starting, but that is an advantage and not otherwise, because they can be destroyed by cultivation before sowing. If the land is full of weed seed, the best thing to do is to start it and kill it. The trouble with dry-plowing probably arises, not from the plowing, but from lack of work enough between the plowing and the sowing. Stubble should often be burned: it depends upon the soil and the rainfall. On a heavy soil with a good rainfall, plowing-in stubble is an addition to the humus of the soil, because conditions favor its reduction to that form, and there is moisture enough to accomplish that and promote also a satisfactory growth of the new crop.

Treatment of Dry-Plowed Land.

We are plowing a piece of light sandy mesa land, dry, which has considerable tarweed and other weeds growing before plowing. Which would be best, to leave the land as it is until the rains come and then harrow, or harrow now? Would the land left without harrowing gather any elements from the air before rain comes! The above land is for oat hay and beans next season.

Roll down the 'tar-weed, if it is tall and likely to be troublesome, and plow in at once so that decay may begin as soon as the land gets moisture from the rain. It would be well to allow the land to lie in that shape, and disc in the seed without disturbing the weeds which have been plowed under. If all this is done early, with plenty of rain coming there is likely to be water enough to settle the soil, decay the weeds, and grow the hay crop. Of course, such practice could not be commenced much later in the season. The land gains practically nothing from the atmosphere by lying in its present condition. If there is any appreciable gain, it would be larger after breaking up as proposed. In dry farming, harrowing or disking should be done immediately after plowing, not to produce a fine surface as for a seed bed, but to settle the soil enough to prevent too free movement of dry air. If your rainfall is ample, the land may be left looser for water-settling.

For a Refractory Soil.

What can I do to soil that dries out and crusts over so hard that it won't permit vegetable growth? A liberal amount of stable manure has been applied, and the land deeply plowed, harrowed and cultivated, but as soon as water gets on it, it forms a deep crust on evaporation. Will guano help, or is sodium nitrate or potash the thing?

None of the things you mention are of any particular use for the specific purpose you describe. Keep on working in stable manure or rotten straw, or any other coarse vegetable matter, when the soil is moist enough for its decay. Plow under all the weeds you can grow, or green barley or rye, and later grow a crop of peas or vetches to plow in green. Keep at this till the pesky stuff gets mellow. If you think the soil is alkaline, use gypsum freely; if not, dose it with lime to the limit of your purse and patience, and put in all the tillage you can whenever the soil breaks well.

More Manure, Water and Cultivation Required.

I have a small place on a hillside, with brown soil about one to two feet deep to hardpan and I am getting rather discouraged, as so many things fail to come up and others grow so very slowly after they are up. A neighbor planted some dahlia roots the same time I did. Only one of mine came up and it is not in bloom yet, while several of his have been blooming for some weeks and are ten times as large in mass of foliage as mine with its lone stalk and one little bud on the top. Peas came up and kept dying at the bottom with blossoms at the top tilt they were four or five feet high, but I never could get enough peas for a mess. Can you help me get this thing right?

Use of stable manure and water freely. Your trouble probably lies either in the lack of plant food or of moisture in the soil. This, of course, is supposing that you cultivate well so that the moisture you use shall not be evaporated and the ground hardened by the process. During the summer a good surface application of stable manure to which water can be applied would be better than to work manure into the soil, which should be done at the beginning of the rainy season. As your soil is so shallow it will be well for you to stand along the side of the plant much of the time with a bucket of water in one hand and a shovel of manure in the other.

Planting Trees in Alkali Soil.

My land contains a considerable quantity of both the black and white alkalies, the upper two feet being a rather heavy, sticky clay, the next three feet below being fine sand, containing more or less alkali, while immediately underneath this sand is a dense black muck in which, summer and winter, is found the ground-water. Do you think the following method of setting trees would be advantageous. Excavate for each tree a hole three feet in diameter and three feet deep. Fill in a layer of three or four inches of coarse hay, forming a lining for the excavation. Then fill the hole with sandy loam in which the tree is to be set. The sandy loam would give the young tree a good start, while the lining of hay would break up the capillary attraction between the filled-in sand and the ground-water in the surrounding alkali-charged soil.

The fresh soil which you put in would before long be impregnated through the surface evaporation of the rising moisture, which your straw lining would not long exclude. The trees would not be permanently satisfactory under such conditions as you describe, though they might grow well at first. It would be interesting, of course, to make a small-scale experiment to demonstrate what would actually occur and it would, perhaps, give you a chance to sell out to a tenderfoot.

Planting in Mud.

Why does ground lose its vitality or its growing qualities when it is plowed or stirred when wet, and does this act in all kinds of soil in the same way? We are planting a fig and olive orchard at the present time, but some were planted when the ground was extremely wet. The holes were dug before the rain and after a heavy rain they started to plant. After placing the trees in the holes they filled them half full with wet dirt, in fact so wet that it was actually slush. What would you advise under the circumstances and what can be done to counteract this? We have not finished filling in the holes since the planting was done, which was about a week ago.

The soil loses its vitality after working when too wet, because it is thrown into bad mechanical (or physical) condition and therefore becomes difficult of root extension and of movement of moisture and air. How easily soil may be thrown into bad mechanical condition depends upon its character. A light sandy loam could be plowed and trees planted as you describe without serious injury perhaps, while such a treatment of a clay would bring a plant into the midst of a soil brick which would cause it to spindle and perhaps to fail outright. The best treatment would consist in keeping the soil around the roots continually moist, yet not too wet. The upper part of the holes should be filled loosely and the ground kept from surface compacting. The maintenance of such a condition during the coming summer will probably allow the trees to overcome the mistake made at their planting, unless the soil should be a tough adobe or other soil which has a disposition to act like cement.

Electro-Agriculture.

Kindly tell me of any one who is working upon the application of electricity to stimulating agricultural growth-especially here on the Coast. A friend who has done some work in this line seeks to interest me. I have seen notices of this work, and have read of Professor Arrhenius stimulating the mental activity of children, etc., but I desire more definite information, if possible. Does the idea seem to you to be feasible?

So far as we know, there has been no local trial of the effect of electric light in stimulating plant growth. Much has been done with it in Europe and in this country. There is much about it in European scientific literature. It is perfectly rational that increased growth should be attained by continuous light in the same way, though in less degree than occurs in the extreme north during the period of the midnight sun. It is known that moonlight, to the extent of its illumination, increases plant growth, and it has been amply demonstrated that light is light, just as heat is heat, irrespective of the source thereof. Of course, the commercial advantage must be sought in the relative amount of increased growth and the selling value of whatever is gained in point of time.

High Hardpan and Low Water.

What detriment is hardpan if 14 inches below the surface and in some places 12 inches? I have been plowing so I could set peach trees, but I have been told that they will not grow. I would like your opinion about it. I intended to blast holes for the trees, and the water is 30 feet from surface. The top soil is red sandy and clay mixed, but it works very easily.

You cannot expect much from trees on such a shallow soil over hardpan without breaking it up, because the soil mass available to the trees is small; also because the shallow surface layer over hard-pan will soon dry out in spite of the best cultivation, because there is no moisture supply from below. If such a soil should be selected for fruit trees at all, the breaking through the hardpan by dynamite or otherwise is desirable, and irrigation will be, probably, indispensable.

Depth of Cultivation.

I would be glad to know whether in cultivating an orchard a light-draft harrow could profitably be used, which cultivates three and a half inches deep? I have used another cultivator, and try to have it go at least seven inches.

A depth of 3 1-2 inches is not satisfactory in orchard cultivation, although there may be some condition under which greater depth would be difficult to obtain because of root injury to trees, which have been encouraged to root near the surface. Both experience and actual determinations of moisture in this State show that cultivation to a depth of 5 inches conserves twice as much moisture in the lower soil as can be saved by a 3-inch depth of cultivation under similar soil conditions and water supply. It is all the better to go 7 inches if young trees have been treated that way from the beginning.

Alfalfa Over Hardpan.

I have land graded for alfalfa and some of the checks are low and water will stand on the low checks in the winter. There is on an average from two to three feet of soil on top of hardpan and hardpan is about two feet thick. Will water drain off the low checks if the hardpan is dynamited, and will this land grow alfalfa with profit?

Yes; much of the hardpan in your district is thin enough and underlaid by permeable strata so that drainage is readily secured by breaking up the hardpan. Standing water on dormant alfalfa is not injurious.

Trees Over High-Water.

Which are the best fruit trees to plant on black adobe soil with water table between 3 and 4 feet from surface? The soil is very rich and productive. The land is leveled for alfalfa also; will the alfalfa disturb the growth of trees?

We would not plant such land to fruit at all, except a family orchard. The fruits most likely to succeed are pears and pecans. On such land alfalfa should not hurt trees unless it is allowed to actually strangle them. The alfalfa may help the trees by pumping out some of the surplus water.

Soil Suitable for Fruits.

I am sending samples of soil in which there are apricots and prunes growing, and ask you to examine it with reference to its suitability for other fruits. Will lemons thrive in this soil?

It is not necessary to have analysis of the soil. If you find by experience that apricot and prune trees are doing well, it is a demonstration of its suitability for the orange, so far as soil is concerned. The same would also be a demonstration for soil suitability for the lemon because the lemon is always grown on orange root. The thing to be determined is whether the temperature conditions suit the lemon and whether you have an irrigation supply available, because citrus fruits, being evergreen, require about fifty per cent more moisture than deciduous fruits, and they are not grown successfully anywhere in this State without irrigation, except, possibly, on land with underflow. The matter to determine then is the surety of suitable temperatures and water supply.

For Blowing Soils.

I am going to dry-sow rye late this fall. I want some leguminous plant to seed with the rye for a wind-break crop, not to plow under. The land varies from heavy loam to blow-sand. I have under consideration sweet clover, burr clover, vetches. I see occasional stray plants of sweet clover (the white-blossomed) growing in the alfalfa on both hard and sandy soil. I read in an Eastern bee journal that sweet clover can be sowed on hard uncultivated land with success. Could I grow it on the hard vacant spots that occur in the alfalfa fields?

You can sow these leguminous plants all along during the earlier part of the rainy season (September to December) except that they will not make a good start in cold ground which does not seem to bother rye much. But on sand you are not likely to get cold, water-logged soil, so you can put in there whenever you like - the earlier the better, however, if you have moisture enough in the soil to sustain the growth as well as start it. We should sow rye and common vetch. Sweet clover will grow anywhere, from a river sandbar to an uncovered upland hardpan, but it will not do much if your vacant spots are caused by alkali.

More Than Dynamite Needed.

I have some peculiar land. People here call it cement. It does not take irrigation water readily, and water will pass over it for a long time and not wet down more than an inch or so. When really wet it can be dipped up with a spoon. Hardpan is down about 24 to 36 inches. I have tried blowing up between the vines with dynamite, and see little difference. Can you suggest anything to loosen up the soil?

You could not reasonably expect dynamite to transform the character of the surface soil except as its rebelliousness might in some cases be wholly due to lack of drainage - in that case blasting the hardpan might work wonders. But you have another problem, viz: to change the physical condition of the surface soil to prevent the particles from running together and cementing. This is to be accomplished by the introduction of coarse particles, preferably of a fibrous character. To do this the free use of rotten straw or stable manure, deeply worked into the soil, and the growth of green crops for plowing under, is a practical suggestion. Such treatment would render your soil mellow, and, in connection with blasting of the hardpan to prevent accumulation of surplus water over it, would accomplish the transformation which you desire. The cost and profit of such a course you can figure out for yourself.

Is Dynamite Needed?

I have an old prune orchard on river bottom lands; soil about 15 or 16 feet deep. Quite a number of trees have died, I presume from old age. I desire to remove them and to replace them with prune trees. I have been advised to use dynamite in preparing the soil for the planting of the new trees.

Whether you need dynamite or not depends upon the condition of the sub-soil. If you are on river flats with an alluvial soil, rather loose to a considerable depth, dynamiting is not necessary. If, by digging, you encounter hardpan, or clay, dynamiting may be very profitable. This matter must be looked into, because the failure of trees on river lands is more often due to their planting over gravel streaks, which too rapidly draw off water and cause the tree to fail for lack of moisture. In such cases dynamite would only aggravate the trouble. Dynamiting should be done in the fall and not in the spring. The land should have a chance to settle and readjust itself by the action of the winter rains; otherwise, your trees may dry out too much next summer.

Improving Heavy Soils.

What is adobe? What kind of plants will grow best in adobe? In this Redwood City I find clay-like soil which looks very dark and heavy. What kind of plants will grow best in this soil?

The term adobe does not mean any particular kind of soil. It is applied locally to clay and clay-loam soils indiscriminately. It generally signifies the heaviest, stickiest, crackingest soil in the vicinity. Most plants will grow well on heavy soils if they are kept from getting too dry and too full of water. This is done by using plenty of stable manure and other coarse stuff to make the soil more friable, which favors aeration, drainage, root extension and plant thrift. Friability is also promoted by the use of lime and by good tillage. The particular soil to which you refer is a black clay loam which can be improved in all the ways stated. It is a good soil for most flowers and vegetables if handled as suggested. You can get hints of what does best by studying your neighbors' earlier plantings.

For a Reclaimed Swamp.

I have land, formerly a pond which dried up in the summer months. It has been thoroughly drained now for several years. The land surrounding it is good fertile soil and produces good crops. On this piece, however, crops come up and look fairly well until about two inches high when they turn yellow and die. Mesquite grass and strawberries seem to be the only crops that will live, and they do not do at all well. Sorrel grows abundantly in the natural state.

Apparently the reclaimed land which you speak of needs liming to overcome the acidity in the soil. Common builders' lime applied at the rate of 1000 pounds to the acre at the beginning of the rainy season ought to make the land much more productive and the soil, at the same time, more friable. Deep plowing with aeration will also help the land, and this treatment can begin at once if the soil is workable. Other additions of lime can be made later as they may be required to make the improvement permanent.

Improving Uncovered Subsoil.

What is the best treatment for spots that have been scraped in leveling for irrigation?

The land can be improved by plowing deeply and turning in stable manure or green alfalfa or any other vegetable matter which may decay, rendering the soil rich in humus and more friable. Of course, it will take some time to accomplish this improvement, and it is necessary that there be moisture enough present to cause the material to decay in order that the improvement may be secured.

Sand for Clay Soils.

Will beach sand do adobe or clay soil any good? It gets hard at times and I thought that if I was to put beach sand in the ground the salt in the sand would do the ground harm.

254

It is certainly desirable to mix sand with heavy soil for the purpose of making it lighter - that is, better drained and more friable and therefore improving it for the growth of plants. Sometimes beach sand contains a good deal of salt, which, however, is readily removed by fresh water, and sand hauled and exposed to the rains rapidly loses any excess of salt it may contain. Probably with such an amount of sand as you are likely to use to mix with your adobe, there is no danger at all from salt. Even if such sand should contain considerable salt, if applied at the beginning of the rainy season it would be so quickly distributed as to not constitute a menace to the growth of plants. The worst adobe can be transformed into a most beautiful garden soil by the application of sand and stable manure.

Plowing from or Towards.

Which is the proper way to plow an orchard? First to plow to the trees and then to plow from them, or to plow from the trees and then to them, and your reasons? I have had many arguments with my neighbor farmers.

There is difference of opinion everywhere as to whether the first plowing should be toward or away from the trees. In places where the soil is pretty heavy and the rainfall is apt to be quite large, plowing toward the trees and opening a dead furrow near the center seems to promote rapid distribution of surplus water. If the rainfall is less and arrangements for deep penetration are more necessary, the plowing can well be away from the trees, so as to direct the water toward the row. It is, of course, exceedingly important in this case, that the land should be worked back before it has a chance to dry out by exposure and this is one of the chief objections to the practice, because one is apt to let the land lie away from the trees, hoping for a late rain which may not come. Whatever theoretical advantages there may be in either of these methods, they can only be secured by the greatest care to avoid the dangers which attend them. This uncertainty is the reason why people so generally disagree as to which is the best practice, and they are right in disagreeing.

Dry Plowing and Sowing.

I dry-plowed my grain field to a depth averaging seven inches; it turned up very rough. I then disked and harrowed it, but it is still very rough. I intended to drill the seed, wait for sufficient rain, and harrow to a satisfactory condition, but have been advised to put no implement on after the drill, as a harrow would spoil the work done by the drill, and a slab or roller would cause the ground to bake. If I wait for rain to work the soil before drilling, it will bring the seeding too late.

You have probably done a pretty good job of dry work. If the land is still too rough for the drill, we should broadcast and harrow again. It is not desirable to harrow after the drill, and to roll or rub is likely to smooth too much, because the land would bake or crust after the heavy rains. This would cause loss of moisture and it is therefore better to leave the surface a little rough. You can roll lightly after the grain is up, if the surface seems to need closing a little.

Artesian Water.

I have a large tract of adobe soil, a black clay top soil. For about five months in the year there is not sufficient water on the place. I have sunk wells in different parts, but with very poor results, the further we went down the drier and harder the soil got. What little water we did obtain was unfit for domestic use. Can you give me an idea as to what might be the result of an artesian well in such soil?

Artesian water has nothing to do with the soils. It is a deeper proposition than that. Artesian water comes from gravel strata overlaid with impervious layers of rock or clay in such a way that water in the gravel is under pressure because the gravel leads up and away to some point where water is poured into it by rain falling or snow melting on mountain or high plateau. As the water cannot get out of this gravel until you punch a hole in its lid, its effort will be to shoot up to something less than the elevation at which it gained entrance to this gravel - as soon as your puncture gives it a chance. Geologists who know the locality may be able to tell you that you have little or no chance, but no one can tell you whether you have a good chance or not until he has tested the matter by boring. The quality of the artesian water is determined by its distant source and the bad water you have found is therefore no indication of the quality of what may be below it. No one should enter an artesian undertaking, except to tap a stratum of known depth, without a long purse. Probably one in a thousand of the bores made into the crust of the earth yields as many gallons of artesian water as gallons of various liquids used in boring it - and yet some of them are good wells to pump from because they pierce other strata carrying water, but not under pressure causing it to rise.

Treatment of Alkali.

I am advised that in some cases alkali may be drained and that in others it is treated with gypsum.

Gypsum is not a cure for alkali, but simply a means of transforming black alkali into white, which is less corrosive and therefore less destructive to plants, but there may be easily too much white alkali present - so much that the land would be made sterile by it. You cannot remove alkali by flooding unless two conditions can be assured: first, that the water itself is free from alkali before application to the land; second, that you underdrain the land at a depth of from three to four feet with tile, so that the fresh water on the surface can flow through the soil into the drains, carrying away from the land the alkali, which it dissolves in its course. To flood land even with fresh water without making arrangements for carrying off the alkali water below, is to increase the alkali on the surface as the water evaporates, and such treatment does land injury rather than benefit. We cannot give you any estimate as to the cost of washing out. It depends altogether upon local conditions: whether you use hand work or machinery for the ditching, and what your water will cost.

Alkali, Gypsum and Shade Trees.

Kindly advise how to apply gypsum, and how much, to heavy, sticky soil, the worst sort of adobe and heavily saturated with alkali. We want to plant shade trees. Eucalyptus and peppers succeed fairly well after once started. Gypsum seems to help, but I don't know how much to use.

The amount of gypsum required to neutralize black alkali depends upon how much black alkali there is to be neutralized, and no definite amount, therefore, can be prescribed beforehand as sufficient without a determination of the amount of alkali. In some experiments gypsum to the amount of thirty tons to the acre or more has been used just for the purpose of seeing how much the land would take, and a fine growth of grain has been secured after using that much gypsum, but that, of course, would be out of the question because the outlay would be more than the land or the crop would be worth.

In the planting of trees at some distance apart, the tree can be protected from destruction and enabled to make a stand in the soil by using gypsum on the spot rather than the treatment of the whole surface. In this way five or ten pounds of gypsum could be used by mixing with the soil to fill a good-sized hole.

258

Distribution of Alkali.

I am told by all the ranchers on the east and south sides of the valley that their wells are excellent. But they all say that on the west side - they are bringing up alkali. One also said that the water level was rising throughout all the valley. Is it safe to depend on this in part, or will the alkali spread over all the valley and the foothills?

It is not unusual to find people who predict the rise of alkali almost anywhere except on their own premises. No one can exactly tell where alkali will go, because no one has complete knowledge of the water movement in underlying strata. Wherever the ground water rises on lower levels because of irrigation on higher levels there is danger of the rising of the alkali, for which the only cure is underdrainage with tile so that this rising water is carried to an outflow and not allowed to approach within three or four feet of the surface. If you have such an outflow and desire to undertake the expense of tiling, you can insure yourself against a serious rise of alkali indefinitely. We do not see, however, how alkali can rise to the higher lands of the valley. Its first effect would be to make lakes or ponds in the lowest parts of the valley, and even then the surrounding mesa lands would not be injured.

Plants Will Tell About Alkali.

Please give information as to the application of gypsum to my soil which is somewhat alkaline. I do not care to have an analysis made of my soil, and believe that you can advise me without it.

If your soil is too alkaline for the growth of plants you can demonstrate that fact by experiment, or if it is capable of being used by the application of gypsum, that also can be determined by experiment and noting the behavior of the same plants afterwards. It is rather a slow process but it is sure enough.

Litmus and Alkali.

Is there any simple soil test for alkali that can be made without a chemical analysis?

You can ascertain the presence of alkali by using red litmus paper, which will be turned blue by the alkali in the soil, if the soil is moist enough. This does not determine the amount of alkali, but the quickness of the turning to the blue color and the depth of the color are both attained when the alkali is very strong. When there is less alkali, the reaction is slower and weaker. This test, however, gives you only a rough idea whether the soil is suitable for growing plants. You can tell that better by the appearance of the plants which you find. Any druggist can furnish the litmus paper, and give you a demonstration of how it acts on contact with alkali.

Using Gypsum for Alkali.

Is it better, to kill the black alkali in the soil with gypsum, just to scatter it over an alkalied spot or to plow the soil first and then use the gypsum? I am going to sow alfalfa.

Use the gypsum after plowing, for it will wet down more quickly, and the gypsum has to be dissolved to act freely. The best way to cure your spot is to run an underdrain into it, if possible, so the rain-water can run through the soil freely and take the alkali with it.

Blasting or Tiling.

In planting trees where hardpan is four feet from the surface is it necessary to blast the hardpan, or is there no benefit derived by the blasting?

If there should be a good available soil under a shallow layer of hardpan, which you say is four feet from the surface, it might be of considerable advantage to bore into the hardpan and explode a dynamite cartridge in it. But if your good soil is really only four feet deep and hardpan continuous below, the blast might cause fissures which would prevent standing water in the upper stratum. If you are sure of four feet of good soil above the hardpan you will have no difficulty in growing good trees, if you get the moisture just right and the hardpan slopes in such a way that surplus moisture will move away. If, however, you have hardpan at different depths on the tract, so that it may really make basins which will hold water, you are likely to have trouble from accumulations of water which will not only prevent the roots extending to the full depths of the soil, but will also cause some trees to die. Such a danger could be removed by draining the soil to a depth of three and a half or four feet with tile, in order to prevent accumulations at any point. This would be expensive perhaps, but you would be sure that you had rendered your four feet of soil safe and available. If you trust to blasting you will have to wait several years for the trees to tell you whether you helped them or not.

Effects of Blasting.

I have land which is underlaid with hardpan two or three feet deep and this in turn is underlaid with sand or sandpan. What I would like to know is whether blasting the holes before setting trees would allow more moisture coming from this sandpan, or, rather, what effect it would have as to moisture.

We do not know. It might make the soil better for the trees by allowing escape for surplus water through previous layers. It might allow the tree to root more deeply for moisture in those strata. It might allow water to rise from such strata if they have water under pressure. It might do other things good or bad, according to conditions prevailing under the hardpan. If you are to irrigate the land the effects would probably be good.

The Sub-soil Plow.

I am contemplating using a sub-soil plow for the purpose of breaking plow-sole on grain land. This is about 4 1/2 inches below the surface and is about 5 inches thick. This soil is comparatively loose and seems to be of good quality. Do you think that the sub-soil plow run low enough to break this plow-sole will benefit the land?

There can be no question about the benefit of breaking up this tight stratum, provided you use a long-tooth harrow or a subsoil packer afterward to reduce the land so that it will not be too open to loss of moisture by too free circulation of air. The best way to treat such a soil would be to use a tractor and plow to a full foot of depth, for this, followed by good harrowing, would disintegrate the hard stuff and commingle it with the loose surface soil and make it somewhat more retentive - doing this when the moisture is just right for disintegration and mixing. If you are not ready to go to this expense, a subsoiler, following the plow with another team, would put your land in better shape for dry farming or for irrigation than it is now. Starting late, however, might give you less crop the first year on such deep working than by shallow plowing if the year's rainfall should be scant. It would, however, be a good start for summer-fallowing and a big crop the next year.

Sour Soil.

What is "sour" soil? Is that the name by which it is commonly known, and what is the treatment for it?

Sour soil is soil in which an acid is developed by plant decay and exclusion of air. The proper treatment is the application of lime, and aeration by open tillage and underdrainage.

Old Plaster for Sour Land.

Can house plaster be used in reclaiming sour ground and how much per acre? The ground produces some sour grass - not a great deal. The plaster is from an old building that is being torn down.

House plaster is desirable as an application to land which is sour. It also adds to the mellowness of land which is hard, because of the sand contained in it. It has always been considered a good dressing for garden land. So far as the correction of sourness goes, it is much less active than fresh lime, but it acts in the same way to a limited extent. It is certainly worth using, providing it does not cost too much for delivery, and can be freely used if the land is heavy and needs friability.

Application of Manure Ashes.

Having recently got a lot of manure plentifully supplied with redwood shavings that had been used with the bedding, and being afraid to use the same in that shape, as it takes such a long time for the wood to rot, I reduced the pile to a heap of ashes. How can it be best applied to ornamental trees and shrubbery in a light gravelly soil?

You have done unwisely in burning the manure. We would have taken the risk of a single use of shavings for the sake of the manurial matter associated with them, and this risk of too much lightening of a gravelly soil would be especially small in connection with deep rooting plants like ornamental trees and shrubbery. You have left merely the skeleton of the manure, and much of that of doubtful solubility, if the temperature ran very high by burning in a mass. You need not be fearful about using these ashes. Scatter or spread them over the ground just as you would have spread the manure, let the rains dissolve and carry down what they can and go on with your usual methods of cultivation.

The Best Fertilizer for Sand.

How can I best fertilize soil that is pure sand?

The best fertilizer for pure sand is well-rotted stable manure, because it not only supplies all kinds of plant food, but increases the humus in the soil, which is exceedingly important in making the sand more retentive of moisture as well as more productive.

Fertilizers in Tree Holes.

Would it be harmful to add 2 or 3 pounds of steamed bone meal to the hole of a young tree just before planting?

There would be no injury, providing you mix it with a considerable amount of soil by digging over the bottom of the hole, but our conviction is that on lands which are good enough for the commercial planting of fruit trees, it is not necessary to stimulate a young tree in this way, but that it is better to postpone the use of fertilizers until the trees come into bearing and show the desirability of more liberal feeding. Of course, if young trees do not make satisfactory growth, they may be stimulated either with some kind of a fertilizer or with a freer use of water, and it is generally the latter that they are chiefly in need of.

Wood Ashes and Tomatoes.

Is there any harm to vegetable growing to dig sufficient of wood ashes in for mellowing heavy soil? My tomato plants grew splendidly this year, but the fruits were all rough and wrinkled. I gave them plenty of horse and poultry manure at planting and plenty of wood ashes and falling leaves of cypress later.

Wood ashes do not mellow a heavy soil. The effect of the potash is to overcome the granular structure and increase compactness. Coal ashes, because they are coarser in particles and devoid of potash, do promote mellowness, and are valuable mechanically on a heavy soil although they do not contain appreciable amounts of plant food. You are overfeeding your tomato plants, probably. The chances are that you had poor seed. There is no best tomato, because you ought to grow early and late kinds: there is also some difference in the behavior of varieties in different places.

Was It the Potash or the Water?

Last year the lye from the prune dipper was turned on the ground near two almond trees which seemed to be dying, and to my surprise they have taken a new lease of life. Hence my conclusion that potash was good for our soil.

Your experience seems to justify the application of potash, surely, but the question still remains, how much good the potash did the trees, and how much they needed the extra water which the waste dips supplied. It would be desirable for you to make another experiment with other trees, applying wood ashes, if you have them, or about four pounds per tree of the potash which you use for dipping, scattering well and working it into the soil after it is moistened by the rains, and not using any more water than the trees ordinarily received from rainfall. After this trial you will be in a position to know whether your trees need potash or irrigation - by comparing with other trees adjacent. Besides are you sure that your lye dip was caustic potash and not caustic soda? The latter has no fertilizing value.

Prunings as Fertilizer.

Is orchard and vineyard brush worth enough as a fertilizer to pay for cutting or breaking and putting back on the land?

We should say not. It takes too much labor to put it in any form to promote decay, and is even then too indestructible. It is also possible that its decay may induce root rot of trees. We should burn the stuff and spread the ashes. Vineyard prunings are more promising because more easily and quickly reduced by decay. Vinecane-hashers have been proposed from time to time, but we do not know anyone who long used them.

Gypsum on Grain Land.

Is there any profit in sowing gypsum on grain land, say on wheat or oat crop? At what stage should it be applied and in what quantity?

It would have a tendency to make the surface more friable and therefore better for moisture retention, and it could be used at the rate of 1000 pounds to the acre, broadcasted before plowing for grain. As our soils are, however, usually well supplied with lime, there is a question whether there would be any profit in the use of gypsum, for, aside from lime, it contains no plant food, although it does act rather energetically upon other coil contents. Gypsum is a tonic and not a fertilizer from that point of view. The best way to satisfy yourself of its effect would be to try a small area, marked so as you could note its behavior as compared with the rest of the field.

Gypsum and Alfalfa.

What is gypsum composed of? Is it detrimental to land in future years? Have the lands of California any black alkali in them? I notice my neighbors who sow gypsum on their alfalfa get a very much better yield of hay than those who do not.

Gypsum is sulphate of lime. It is not detrimental to the land in after years except that its action is to render immediately available other plant foods and this may render the land poorer - not by the addition of anything that is injurious but by the quicker using up of plant food which it already contains. Black alkali is very common in California in alkali lands. In lands which show their quality by good cropping, there is no reason to apprehend black alkali nor to use gypsum to prevent its occurrence. The use of gypsum does stimulate the growth of alfalfa and makes its product greater just as you observe in the experience of your neighbors, but the more they use up the land now the less they will have later, unless they resort to regular fertilization to restore what has been exhausted. But even that may be a good business proposition.

What Gypsum Does.

I intend to fertilize alfalfa and should like to know about gypsum. I have heard it stimulates the growth temporarily but in three or four years hurts the land. I have heavy land.

The functions of gypsum are: (a) to supply lime when the soil lacks it; (b) to make a heavy soil more mellow, and (c) to act upon other soil substances to render them more available for plant food. These are some of the soil aspects of gypsum; it may have plant aspects also. It is too much to say that gypsum hurts the land; it does, however, help the plant to more quickly exhaust its fertility, and in this respect is not like the direct plant foods which comprise the true fertilizers - one of which gypsum is not. It might be best for your pocketbook and for the mechanical condition of the soil to use it, but do not think that it is maintaining the fertility of the land (a service which we expect from the true fertilizers) except as it may supply a possible deficiency of lime.

How Much Gypsum?

How much per acre, how frequently and what seasons of the year are the best time to apply gypsum?

Of gypsum on alkali, we should begin at the rate of one ton to the acre and repeat the application as frequently as necessary to achieve the desired result. If the alkali was quite strong we would use twice as much. Without reference to an alkaline condition in the soil, and to give heavy soil a more friable character, which promotes cultivation, aeration, etc., and, therefore, ministers to more successful production, half a ton to the acre can be used, applications to be repeated as conditions seem to warrant it.

Wood Ashes in the Garden.

There is available in my neighborhood a free supply of wood ashes. Can you tell me how best to distribute the same in a garden (flowers and garden truck), and what, if any, treatment is to be given the ashes for the best results.

Wood ashes long exposed to rain lose most of their valuable contents, and leached ashes are only of small value. If they are fresh ashes or ashes which have been kept dry, they are chiefly valuable for potash, which is good in its way, but not all that a plant needs. If, however, your soil is shy of potash, the use of ashes will notably improve growth if not applied in excess in the caustic form in which it occurs in the ashes. They require no treatment. Spread, say, a quarter of an inch thickness all over the ground and dig in deeply. It may also help you by destruction of wire worms and other ground pests.

Coal Ashes in the Garden.

What is the effect of coal ashes on the red clay soil of Redlands or wood and coal ashes combined?

Coal ashes are exceedingly desirable upon clay land because their mechanical mixture with the fine particles of the clay renders the soil more friable, permeable and better adapted to the growth of most plants. Coal ashes, however, possess no fertilizing value - their action is merely mechanical. The wood ashes which may be combined with them are desirable as a source of potash which most plants require.

Liming a Chicken Yard.

I have a small family orchard of half an acre, fenced in as a chicken yard, the soil of which has become very foul. When would be the best time to apply lime and how much?

Put on 500 pounds of lime and plow under as soon as you can - that is, spread the lime just before the plowing, with a shower or two on the lime before plowing, if the weather runs that way.

Poultry Manure.

Give directions for using chicken manure. For use of young trees, is there any difference in treatment of deciduous and citrus trees? For use in the vegetable garden and the flower garden, what should be mixed with it and in what proportions? So many people say poultry manure is so strong, I am afraid to use it.

It is a fact that poultry manure, free from earth, contains even as high as four times as much plant food as ordinary stable manure. It is, therefore, to be used with proportional care, so that the plants shall not receive too much, and particularly so that there may not be too much collected in one place. Probably the best way to guard against this is to thoroughly mix the manure with three or four times its bulk of ordinary garden soil and then use this mixture at about the same rate you would stable manure. If you do not desire to go to all this trouble, make an even scattering of the manure and work it into the soil. There is no reason to fear the material; simply guard against the unwise use of it. It is good for all the plants which you mention; in fact, for any plant grown, provided it is sparingly and evenly distributed.

It should be pulverized so that there shall not be lumps and masses in the same place for fear of root injury. Of course, the strength depends upon how much earth is gathered up with the manure. Sometimes there is so much waste material that it can be handled just as ordinary farm manure is.

We should not use over 20 pounds of clean droppings to a young tree and should mix it with the soil for a considerable distance around the tree. Old bearing trees might stand two or three tons to the acre if distributed all over the ground. The material contains everything that is necessary for the growth of the tree and formation of the fruit.

Ashes and Poultry Manure.

It is said that ashes mixed with chicken manure is not good. I use ashes altogether on the drop boards because I can keep the boards cleaner. The refuse is then scattered around the fruit trees.

Wood ashes and lime should never be used as you propose, because they set free the nitrogen compounds which are the most valuable content of poultry manures. This action is conditioned largely upon the presence of moisture, and if the droppings are kept dry and hurried into the soil the loss is lessened. Coal ashes, on the other hand, are a thoroughly good absorbent when the coal burns to a fine ash or is sifted. They do not act as wood ashes do, because they do not contain soluble alkali. They also have a good mellowing effect on heavy soil.

Caustic Lime Not a Good Absorbent.

Would air-slackened lime be suitable to sprinkle over the dropping boards in hen houses?

Gypsum is greatly superior to air-slacked lime for the hen houses, as it has every beneficial effect of the latter, while the air-slacked lime will set free much of the fertilizing value of the manure, which the gypsum will not do.

Too Much Chicken Manure for Young Trees.

I have peach trees and apple trees, 3 to 6 years old, that are very thrifty but grow only wood. The soil was poor when planting, and I have put on plenty of sweepings from the chicken-yards. I suppose that is the cause of the trouble.

Undoubtedly you have overmanured your soil with chicken manure, which is a very strong fertilizer and should only be used in limited quantities. In order to counteract any acidity or ill effects which have been produced by its excessive application, it would be desirable for you to apply about 500 to 1000 pounds per acre of common builders' lime at the beginning of the rainy season, working it into the soil with the fall or early winter plowing. Do not cut back the tree during the dormant season, although, of course, you may have to remove surplus or interfering branches for the sake of shaping the tree. Winter pruning induces a greater wood growth during the following summer; therefore, it should be avoided under such conditions as you describe. Having adopted such a policy, there is nothing for you to do but to wait for the trees to slow down and assume a normal bearing habit proper for their ages. Summer pruning is an offset for excessive wood growth.

Suburban Wastes.

We keep a cow and poultry and have a dry-earth toilet. We have been burying the manure in the little garden spot or along by the fences or spreading it out on the alfalfa before it is rotted, but do not get good results. How shall we apply it to get the best results ? We have a town ordinance against leaving it in piles to rot.

You can compost it in a tight bin made of planks, and using enough water to prevent too rapid fermentation and loss of valuable ingredients. During the dry season you can probably use enough dry earth or road dust to render the material inoffensive, and you can also distribute it then without undesirable results.

Composting Garden Wastes.

You recommend making a compost of all scrapings, garbage, weeds, etc. Is there any danger in having this in a pit near the house?

If you desire to put garden wastes, including manure, into a pit, the only objection would be the heavy work of digging it out again. If you allow waste water from the house to run into the pit, there would probably be not enough dry material to absorb it, and the pit would be not only objectionable on account of odors, but possibly dangerous to health. The water would also prevent decomposition, because of exclusion of air. At the same time, enough moisture to promote slow decomposition is essential. It is usually more convenient to compost garden wastes on the surface of the ground, enclosing them with a plank retainer, because moisture can easily be applied with a hose, as desirable, the material can be occasionally forked over to promote decay, and the heavy work of digging material out of a pit is avoided. Such a collection is neither offensive nor dangerous if handled right.

Composting Manure.

Will the dry barnyard manure, when heaped up and dampened with water, make a valuable fertilizer?

For garden use, dry manure in heaps should be dampened with water from time to time so as to prevent too active fermentation. Of course, water should not be supplied so freely as to cause a leaching of the pile. It is also desirable that the material should be forked over from time to time to distribute moisture and promote decay. When this is done a thoroughly first-class fertilizer is produced.

Barnyard Manure and Alkali.

In spots my land is hard and has some black alkali. Will barnyard manure help the hard land if cultivated in?

Use stable manure because that would not only furnish nitrogen, if your plants need any more, but it would add coarse material and ultimately humus which would overcome the tendency of your soil to become compact and thus concentrate alkali near the surface by evaporation. Mellow the soil, increase the humus, make water movement freer and good cultivation easier and alkali will become weaker by distribution through a greater mass of the soil and may be too weak at any point to be troublesome, unless you have too much to start with. Put on manure at the beginning of the rainy season and plow it under, with all the green stuff which grows upon it, during the winter or early spring.

Stable Manure and Bean Straw.

What are the approximate contents of common stable manure; also, how much of the above is contained in bean straw?

The composition of mixed stable manure is given as containing in one ton: Nitrogen, 10 pounds; phosphoric acid, 5 pounds; potash, 10 pounds. The constituents of bean straw in one ton, are given as: Nitrogen, 28 pounds; phosphoric acid, 6 pounds; potash, 38 pounds; Of course, a large part of the difference in composition is due to the excessive amount of moisture which ordinary stable manure contains. Air dried stable manure, such as is found in a California corral, would have much higher fertilizing value than such moist manure as an Eastern chemist would be likely to handle.

Roofing a Manure Pit.

Is it necessary to roof a manure pit, if the pit is tight so that all rain on manure is caught in the liquid manure and nothing is lost?

To secure satisfactory composting of stable manures in a pit it is necessary to be able to regulate the moisture of the mass. If it becomes too dry, too rapid fermentation takes place and the material is destroyed by what is called fire-fanging. If too much liquid enters the pit, so that the material is submerged, the air is excluded and fermentation stops. For these reasons it is necessary that a pit in the region of large rainfall be covered, and water be used from a hose or other source of supply in just sufficient quantity to keep the material right for slow fermentation. How much water should be added to bring the moisture to a right condition depends upon how much liquid waste runs into the pit, and where water is used for cleaning a stable care has to be taken that the pit is not submerged. Success with a pit is, therefore, conditioned on the amount of moisture admitted, and this cannot be controlled unless the pit has a cover fit to shed rainfall. Of course, it may be adjustable so that some rainfall may be admitted as may be desirable.

Value of Animals in Manure.

In the operation of our fruit and dairy ranch we have the manure from some forty head of horses and cattle, which is distributed over the place. We cut our alfalfa and feed it and do very little pasturing. In order to give our dairy the proper credit, we would kindly ask what you consider a fair price for the manure of a cow for one year. Also what would the manure from a horse for one year be worth?

A compilation of a considerable number of weighings, analyses and valuations in Europe, cited by Prof. Roberts in his book on the "Fertility of the Land," gives an average value of the voidings of a cow for a year as $32.25 and of a horse at $24.06. This is based, of course, upon the collection and saving of all excrements which is never secured except in careful experimentation. The value of manure depends upon the quality of the feed. In two experiments, considered a safe substitute for the straw, apart from the fact that the gave a value in manure of $1 per ton of hay fed; cows fed on clover and bran gave value in manure of 3.80 per ton of mixed feed. Your alfalfa feeding would approach the higher value. You will have to make an estimate from the above data to serve your purpose and you can figure it either by the number of animals or by the tonnage of the feed.

Value of Fresh and Dry Manure.

What is the relative value of the weekly or semi-weekly corral scrapings which are tramped fine and air-dried; and of the fresh, wet manure from the stable? I do not understand that the latter has appreciable water added, and the amount of sand in the corral scrapings would be small.

Fresh, mixed animal manure is usually calculated to contain about 75 per cent of water. Manure which has been quickly dried, without fermentation and without leaching by rains, may be worth four or five times as much per ton. Nothing, however, short of analysis would determine the value of any particular lot, for that depends somewhat upon the way the animals are fed, as well as upon the moisture content.

Shavings in Stable Manure.

Is barnyard fertilizer containing shavings instead of straw, desirable?

Barnyard manure containing shavings is chiefly objectionable because of the amount of inert material. The shavings are exceedingly slow to decompose, and in light soil in considerable quantities would cause a serious loss of moisture. If applied, on the other hand, to a heavy soil and accompanied by sufficient irrigation water, the effect of making the soil more friable might be very desirable. It depends then upon circumstances whether shavings can be concited by Prof. Snyder in his "Soils and Fertilizers," cows fed on hay straw is more valuable not only because more easily decomposed, but because its content of plant food is greater.

Handling Grape Pomace.

In the case of grape pomace, would not the large value shown by analysis be chiefly in the seeds? My observation is that these are exceedingly slow to became available in the soil. Would composting break down the shell of the seed?

Grape pomace is slowly available because of the slow disintegration you mention. It could be hastened by drying and grinding, but we doubt if this or other treatment would return its cost. Decay by moisture promoted by composting with manure, kept at a low temperature by continuous moisture would render it sooner available, but this would involve labor which, at our wage rates, would probably make the material cost more than it is worth. This is probably a cost in which time is cheaper than money.

Sheep and Goat Manure.

I can buy goat manure from an inclosure where this is deposited to an amount of about five carloads. Will goat manure be of great value in fertilizing an orchard? If so, how much of it should be spread an an acre?

Accumulations of sheep and goat manure in a dry situation, that is, where not leached out by heavy rainfall, have been found to run as high as $13 per ton in fertilizing constituents. The average would, however, be not above $7.50, and would depend not only upon the unleached condition of the material but upon the amount of sand mixed with it. If it is in a situation where sand blows very freely, it might not be worth over $4 or $5 per ton, possibly not that much. You have, therefore, to deal with a condition largely unknown. So far as its fertilizing quality goes, however, it is freely available and directly calculated to stimulate the growth of plants, and probably four or five tons could be used to the acre without injury if well distributed over the surface of the land. Application can be made at any time of the year, for the drying will not injure it. It will not, however, become available until the soil is sufficiently moist to carry its contents to the roots of the plants. Under ordinary conditions in California, application should be made just before the beginning of the rainy season.

Hog Manure and Potatoes.

What is the fertilizing value of hog manure, and also what is the best fertilizer to use for potatoes? Our potatoes are planted early in January.

Hog manure is rather a rank and strong fertilizer, usually very rich, although the quality of it depends upon how well the hogs have been fed - that from grain-fed hogs being notably better. The valuation of hog manure ranges from $2.50 to $3.25 per ton, according to the feeding as noted, while ordinary stable manure may be worth from $2 to $2.75 per ton. It is not a good idea to apply these organic manures directly for the growth of potatoes. It is better to apply them to the land for the growth of a grain or forage crop, plowing in the stubble and using the land for potatoes the following year. If you wish to fertilize directly for potatoes, the use of a commercial fertilizer containing a good amount of potash would be a better proposition.

Fertilizer for Sweet Potatoes and Melons.

I have sandy soil that has been used for sweet potatoes until it is worn out for that crop, and would like your advice as to the best fertilizer to use. Also, what fertilizer would be best for melons on land that has been planted to melons for the past three years?

There is not much difference in the plant food required by the two crops you mention, but both evidently need a freshened soil and an increase of humus. We should apply a half ton to the acre of a complete fertilizer, of which any dealer can give you descriptions and prices. If you wish to do a good job, start a growth of peas or vetches or burr clover, and sow the fertilizer evenly with the seed. Plow the growth under in February and roll (as the soil is sandy) to close down and promote the decay of the green stuff, which ought to be so well accomplished by the date that it is safe to plant sweet potatoes or melons that it will give no trouble in summer cultivation.

An Abuse of Grape Pomace.

I got in an argument with a neighbor of mine who stated that grape pomace is not a fertilizer. Is it so? My neighbor says that two years ago he had two apricot trees in his yard, and they were fine bearing and healthy trees. After making his wine he put the pomace on the ground and they died. Could that be the cause?

Yes, probably. He used too much fresh pomace and the resulting fermentation of its products may have killed the trees. But grape pomace, after going through fermentation and in the process of decay, makes humus in addition to giving potash and other desirable substances to the soil.

Manuring Vineyard.

Does barnyard manure have any injurious effect on the vines if applied on my vineyard? One of my neighbors claims barnyard manure burned his vines so he got no crop wherever he spread the manure, and nothing would now induce him to use it again.

Barnyard manure can be safely used in a vineyard at the beginning of the rainy season, working it in with the plowing, but not using too much. Wine grapes are sometimes injuriously affected in flavor by the use of such fertilizer, but the growth of the vine itself can be stimulated by the rational use of it. Your neighbor apparently either used too much or made the application at the beginning of the dry season or made some other mistake.

Bones for Grape Vines.

I am going to plant out some grape vines, and would like to know if it is a good plan to put old bones, broken up fine, into the holes when planting.

Yes, if you do not use too much and it is mixed with earth, a little beyond the touch of the roots at planting. You do not need to finely break the bones. The roots will take care of that. But do not put in too much coarse stuff, for fear of causing too rapid drainage.

Reviving Blighted Trees.

I have a couple of apple trees here that were hurt by the pear blight three years ago and were cut back since then; they come out each year, but the leaves curl up, and they do not do anything. I would like to know if putting any fertilizer around them would help them to put out their leaves, and if so what I should use?

Put some stable manure on the top of the soil around your trees now so that the rains may reach the contents of the soil, then later in the season dig the manure into the soil. Apply water during the summer time and this will encourage the trees to grow, if there is any vigor remaining in them. This treatment, however, will not protect them from the blight.

Fertilizing Pear Orchard.

I have pear trees 15 years old which have fruited heavily for years and have never been fertilized. What is the best fertilizer for the soil which is heavy, and when is the best time to apply it? I intend planting rye to plow under in the spring, but thought possibly the fertilizer should be applied first.

If you have stable manure available, nothing could be better for the feeding of the trees and for its mellowing effect upon your heavy soil. Application can be made at once, to be worked into the land when the rye is sown. It will help the trees and give you more rye which in the end will help the trees. If you have no stable manure available, what is called by the dealers a "complete fertilizer" for orchard purposes is what you should use and apply it when you work the land for rye.

Fertilizing Olives.

What is the best means of fertilizing an olive orchard? My orchard gives me a perfect quality of oil, but a poor quantity. My soil is dry calcareous, red and gray, and is very thin in places, therefore, it lacks moisture.

An olive orchard can be fertilized with stable manure or with a "complete fertilizer," or with the special brands of different manufacturers of special fruit fertilizers. But you must be sure that your trees do not need moisture more than they need fertilizers, for without adequate moisture fertilizers cannot do their best work. The increase of the humus content of the soil, either secured by stable manure or by the plowing under of winter-grown cover crops, is desirable, as they not only give the trees more plant food, but make the soil also more retentive of moisture. You will have to experiment along this line to see just what is best for your trees.

Consult the Trees.

Can I send you a little soil out of my one-year-old pear orchard so that you can advise me what I can do to improve its fertility. The trees are fairly thrifty, but as fruit growing is my pleasure I wish to make it a model orchard and add whatever it requires of nitrogen, humus, etc., immediately so as to increase the growth for this summer. Next winter I intend to put manure around them and cultivate about every other month.

Careful experimenting with fertilizers will teach you more than analysis would do, because the behavior of the tree under various conditions tells you more than a chemist possibly could. Besides, we are of the conviction that on good soils young fruit trees should not be pushed beyond the growth which they would naturally make with a regular and adequate moisture supply. Be careful about using fertilizers on young trees, either in the summer or in the winter. When they come to bearing age and yield large crops of fruit, that is another question. Any California soil which will not grow young fruit trees thriftily should not be used for orchard purposes unless an amateur desires to grow trees on a picturesque lot of rocks or sand.

Results of Fertilizing Olives.

We have 100 acres in olives about six miles northeast of Rialto in San Bernardino county. In 1908 we got about five tons from the 100 acres. We began fertilizing and cultivating in 1909, and have put on the 100 acres about the same amount of fertilizer each year. In 1909 we got 15 tons; in 1910, 116 tons, and 1911 is estimated at 325 to 350 tons.

It is important that your olive trees are responding to good treatment and fertilization. Unfortunately, that does not seem to be always the case and a good many olive trees have been made into firewood because nothing seemed to bring them into satisfactory bearing. Good bearing olive trees are now among the very best of our horticultural properties, while non-bearing olive trees are worth about $7 a cord for fire wood.

Nursery Fertilizers.

I have light sandy loam, well drained. It has been in blackberries, and I now have it planted to nursery fruit tree stock. I have given it this spring two applications of nitrate of soda, but no other fertilizer. Will the nitrate act alone, or must I apply also the phosphate and potash to get results?

Nitrate of soda will act alone and will stimulate growth, and there are cases in which there is enough phosphate and potash already in the soil to act with it. Usually, however, it is customary to use a complete fertilizer containing phosphate and potash as well as nitrogen, in order that the plant may be more roundly supplied and promoted, and one would be a little safer in using that sort of fertilizer than in relying upon the nitrate of soda alone. You will, of course, be careful not to use these fertilizers in too large amounts, for nitrate of soda is especially dangerous if used in excess.

Almond Hulls and Sawdust.

Is there any fertilizing value in the hulls of almonds? Would pine sawdust from the lumber mills be a good substance to mix in and plow under in a three-acre adobe patch in order to loosen and lighten the soil for truck gardening?

Almond hulls have considerable fertilizing value, but they are slow to decompose, and, therefore, may be a long time unused by the plant. They also have a good feeding value for stock, and if you can expose them in the corral so the stock can eat as they like, this is the best way to get them into fertilizing form. If they can be cheaply ground their availability as a fertilizer would, of course, be quickened. Redwood sawdust is better than pine sawdust, but any kind of sawdust can be made to serve a good purpose in mellowing heavy soils if not used to excess and if there is plenty of moisture to promote decay.

Fertilizing Fruit Trees.

I have an orchard of prunes, apricots and cherries, which has been bearing since some 30 years ago, without fertilization, except possibly muddy sediment from occasional irrigations of mountain streams. Various people are advocating the use of nitrates and other fertilizers. Should I have samples of this earth analyzed in order to ascertain what the soil most needs?

To find out whether your trees need fertilization, study the tree and the product and do not depend upon chemical analysis of the soil. If your trees are growing thriftily and have sufficiently goodsized leaves of good color, and if fruit of good size and quality is obtained, it is not necesssary to think of fertilization. If the trees are not satisfactory in all these respects, the first thing to do is to determine whether they have moisture enough during the later part of the summer. This should be determined by digging or boring to a depth or three or four feet in July or August. The subsoil should be reasonably moist in order to sustain the tree during the late summer and early fall when strong fruit buds for the coming year will be finished. If you are sure the moisture supply is ample, then fertilization either with stable manure or with commercial fertilizers containing especially nitrates and phosphates should be undertaken experimentally, in accordance with suggestions for application made to you by dealers in these articles, who are usually well informed by observation. When you have the tree to advise you of the condition of the soil, you do not need a chemist, although if the tree manifests serious distress and is unable to make satisfactory growth the suggestions of a chemist may be very helpful.

Fertilizing Oranges.

What is the general and what do you consider the ideal, manuring, and when applied for orange trees from 15 to 12 years old under irrigation? I use about 2 cwt. each of superphosphate, nitrate of soda and sulphate of potash per acre, but am dissatisfied with my yields as compared with yours in California.

There is not only no standard for fertilizing orange trees, but there is no "ideal" which might be considered as a basis for a standard. All growers who are awake to the necessity of doing something for bearing trees, try all things and hold fast to what (they think) is good. Practically none of them has any enduring conviction or demonstration as to what is good, but they keep on trying. There is, however, one clear and enduring conviction, and that is, that continuous fertilizing must be done for profit, and our best growers are using the same materials you mention in considerably larger amounts than you apply, and use also other forms of nitrogenous fertilizers. The amounts of superphosphate and nitrate which you use would be considered homeopathic treatment by our growers.

Cow Stable Drainage for Fruit.

I have been told that the drainings from a cow barn make an excellent fertilizer for orange and lemon trees, in fact, anywhere on plants where manure is considered beneficial.

The drainage from a cow barn is excellent for fertilizing almost any crop unless it is used in too large quantity. If it should be combined with a considerable amount of water used for cleaning out the stable, it would be excellent for the irrigation of all kinds of fruit trees. Care should be taken, however, not to oversaturate the ground, which would be the case if the washing of the stable was allowed to run continuously alongside a single row of trees. The water should be changed from row to row in succession, cultivating the ground meantime to promote aeration and to prevent too great compacting of the soil.

Seed Farm Refuse as a Fertilizer.

Would cleanings from sweet peas or all kinds of seeds grown on a seed farm be of any value as a fertilizer on sandy loam soil for an orchard? This has been in a pile for three years or more, and I can get it for the hauling. There are a hundred loads or more of it and not very far to haul.

It would be worth more on a heavy soil, because the danger of drying out would be less and the surety of reduction to humus greater. To get the highest value from such stuff it should be composted with water and turning in heaps, but that would occasion expense beyond value probably, unless it could be composted with manure for market garden purposes. The hauling might be good work for idle teams. Spread the stuff rather thinly to be covered in with fall plowing, so that its decay could be promoted during the rainy season.

Slow Stuff as a Fertilizer.

How can we use sawdust and shavings from our high school shop so as to combine it with street sweepings, lawn cuttings, etc., and insure ready decay without objectionable features?

Do not mix sawdust and shavings with lawn clippings and street sweepings, because of the great difference in susceptibility to decay. The lawn clippings and street sweepings, which would contain considerable horse manure, would be readily transformed into a good fertilizer by composting. Such treatment, however, would have no appreciable effect upon sawdust or shavings for a considerable period of time, and they would still be too coarse in their character to be of any value unless you have to deal with heavy clay soil, and in that case the sawdust and fine shavings might be dug in at once and trusted to decay slowly in the soil, at the same time improving its friability by their coarser particles. If, however, you are dealing with light sandy loam, such coarse material would cause too rapid drying out and injure the plant, which might be benefited by lawn clippings and street sweepings. The best way to get rid of the sawdust and shavings is to set up an altar, such as we have in our own backyard - a piece of an old boiler about two feet in diameter and two and a half feet high, in which we currently burn all rubbish which is not available for quick composting into a fertilizer.

Lime on Sandy Soil.

Do you think 500 pounds of lime per acre would help a sandy soil which has not been enriched by pasturing or legumes? Of course, we would not apply the lime until next fall before plowing.

Lime is not usually called for in a sandy soil, which probably requires direct fertilizing with stable or commercial fertilizers.

Lime on Alfalfa.

What effect does putting lime on land have in holding moisture? Also, will it pay to put it on a large field of alfalfa? The land is adobe. I can get slaked lime for the hauling, distance being about five miles.

The lime will make the land more friable and, therefore, less disposed to bake and lose moisture by evaporation. Alfalfa is hungry for lime and is generally advanced by the application of it.

Fertilizing Alfalfa.

Can new cow manure be put on alfalfa? Is not the best way to use the above as a fertilizer in form of liquid being run from barn via pipes to a settling-tank and from there via irrigation ditches to the land to be irrigated? What is the best way to get rid of cow manure so as to keep a barn sanitary and the place free from stench?

Cow manure can be used to advantage on alfalfa. Corrals can be cleaned up and the manure spread at the beginning of the rainy season. During the winter the manure can be spread as it is produced and very good results will be noticed in the growth during the following summer. It is perfectly rational for you to use the liquid fertilizer as you propose in connection with irrigation water, but this is not generally done because of the cost of the outfit and the labor of handling the material in that way. The best way to keep a barn sanitary is to keep it clean, removing all the waste matter to a considerable distance daily, allowing nothing to accumulate, and have the stable drainage arranged so that the stable can be frequently flushed out into good drainage outlets, carrying the water to grass or alfalfa land if possible.

Fertilizing Corn.

We are going to plant about 20 acres to corn on a sidehill and intend to put some fertilizer on, but want to give it to the corn only. Would it be a good plan, after we have marked out our rows, to scatter some fertilizer in these marks and put the corn right on top of it?

We take it you ask about the use of a readily soluble commercial fertilizer. If so, you can do as you propose, being careful not to use too much. The operation of planting will distribute the fertilizer through enough soil if the application is not too heavy. The effect will depend something upon what showers you get after planting.

Scrap Iron as a Fertilizer.

Is cast or other iron in small pieces plowed into the land of any benefit to trees as a fertilizer? If so, what would be the value as such per 100 pounds? Junk dealers sometimes offer 25 cents per 100 pounds. If it has any value as a fertilizer, I am satisfied it must be worth four times that price. We pay three cents a pound for sulphate of iron as a fertilizer. Of course, it is a salt and dissolves quickly, therefore, I believe cast iron, even if it works slowly, has some value, and at the same time farmers can clean up and get rid of a lot of rubbish.

In most cases the California soils are sufficiently supplied with iron by nature. Iron scraps have a little and remote value because they are so slowly available by the process of rust disintegration. It might, therefore, be worth while for farmers to bury such scrap iron as accumulates on the place below the reach of the cultivating tools. But it would not be profitable to buy iron scraps at junk dealers' price, nor would it be profitable to haul this material any long distance, even if it could be had for nothing.

Kelp as a Fertilizer.

Are there ill effects from using sea kelp as a fertilizer for orange trees?

There is no ill effect. Sea kelp has been dragged from the beaches at low tide, partly dried and used, for centuries perhaps, as field fertilizer for all sorts of crops in Europe, and for decades, to some extent, on the New England coast. The dangerous substance in it would seem to indicate that that is not present in sufficient quantity to cause trouble. The great difficulty lies in securing and transporting the substance, for less than its fertilizing equivalent can be obtained by purchase of other more concentrated manures.

Applying Thomas Phosphate.

When is the best time to apply Thomas phosphate slag on orchard land?

As Thomas phosphate is slowly soluble, it can be applied at any time during the rainy season without danger of loss, and for the same fact, it should be applied early during the rainy season in order to be available to trees during the following summer's growth. It ought, perhaps, to be added that other forms of phosphate have largely displaced slag during the last few years in the United States, other forms being more available.

Sugar Factory Lime for Fertilizing.

Is the lime from a sugar factory a good fertilizer for either oranges or walnuts; if so, about what amount to the acre would you recommend?

If your land needs lime or if it is heavy and needs to be more friable, or if you have reason to think that it may be soured by exclusion of air or by excessive use of fermenting manures, the refuse lime you speak of will do as a corrective just as other lime does, though, perhaps, not so actively. Beyond that there is nothing of great value in it. You can use two or three applications of 500 pounds to the acre without overdoing it - if your land needs it at all.

Nitrate With Stable Manure.

I am going to plant about 2000 plants of rhubarb. I intend to put some cow and horse manure under the plants as a fertilizer, but I do not think I will have enough for all the plants, so I bought some nitrate of lime, with the intention of mixing the cow and horse manure with the lime nitrate, which I thought would allow me to spread the manure much thinner and I could cover more surface. Now I am not sure but the nitrate of lime will burn the manure if mixed with it.

You can mix either nitrate of lime or nitrate of soda with the stable manure as you propose; in fact, it is frequently done. These nitrates are neutral salts and do not act on manure as caustic lime or wood ashes would do. They are quite content to keep along without kicking their neighbors. But, of course, the more nitrate you add the more careful you must be about using too much of the mixture, and as for putting manure under any plant, at spring planting particular, it is dangerous business.

Nitrate of Soda.

How shall I apply nitrate of soda as fertilizer for roses and other flowers and lawns during the summer months?

One has to be very careful in the use of nitrate of soda not to use too much and not to apply it unevenly, so that too much is brought in contact with the roots of particular plants. From one to two hundred pounds an acre evenly distributed is the usual prescription for nitrate of soda, although in the case of bearing orange trees considerably larger amounts have been successfully used. This would be at the rate of about one ounce to one square yard of surface. It would be a safe application to begin with and could be increased a little on the basis of observation of results. Of course, the application should be accompanied by copious irrigation in order to dissolve and distribute the substance.

Fertilizing Strawberries.

I have half an acre of strawberries which will fruit their second season this spring, and half an acre set last month. I had intended to use nitrate of soda on them, but was talking to a friend who told me it would kill my soil. That the first year it would produce an enormous crop and the next year I couldn't raise anything. Which would be better to use here, stable manure or commercial fertilizer?

It is true that nitrate of soda is a stimulant of plants, and by rendering soil fertility immediately available may seem to reduce the supply later, and yet it is a most available forcing fertilizer if used with great caution, not over 200 pounds to the acre evenly scattered over the whole surface or a less amount, of course, if confined to particular areas. If used in excess it may actually kill the plants. Still nitrate of soda is being used actively and intelligently by nearly all growers of plants and must be counted on the whole a valuable agency. If you can get stable manure, nothing is better as a complete plant food. Application to strawberries must be made at the close of the season, rubbish scraped away and manure applied and allowed to stand on the surface during the early rains, being worked into the soil during the rainy season. If the soil is light, sandy loam, too much coarse material must be avoided. Therefore, well-rotted manure is important on such soils while on a heavy soil coarser material may be used to advantage if applied early in the rainy season. If you have no well-rotted manure, a complete commercial fertilizer will give best results.

Late Applications of Nitrate.

I have some prune trees which blossomed some time ago and the prunes are already set, and of small size. Would you recommend me to use an application of, say 100 pounds per acre of nitrate of soda, applied immediately, or is it a little too late in the season to get the desired result?

It would be perfectly safe to use 100 pounds of nitrate of soda to the acre well distributed now; in fact, you could safely use twice as much, but we doubt if you would get any benefit from it unless you should irrigate, for there is no reason to expect showers that would have penetrating powers enough to carry the nitrate any appreciable distance into the soil. Of course, the nitrate could be plowed or cultivated in to a considerable depth, but that would probably result in losing moisture by deep opening or turning, which would do more harm than any gain which the nitrate produces, if it were to become available. Our judgment would be, then, that it is too late for any benefit to accrue unless the land can be irrigated.

Charcoal is a Medicine, Not a Food.

Recently a lumberyard burned, leaving quite a quantity of charcoal. I have a lot 50 x 150 feet in rhubarb. Would the charcoal be of any service on that lot as a fertilizer? I now have it well fertilized with horse manure, but would like to use the charcoal if it would be of any material assistance to the plants.

Charcoal is of no value as a fertilizer. It is practically indestructible in the soil. In fact, they are digging up now charcoal in the graves of ancient Egyptians, who departed this life five thousand years ago. Charcoal has corrective influence in absorbing some substances which might make the soil sour or otherwise inhospitable to plants. It has been found desirable sometimes to mix a certain amount of charcoal with soil used in potting plants for the purpose of preventing such trouble. The only way to make your charcoal of any value as a fertilizer would be to set it on fire again and maintain the burning until it was reduced to ashes, which are a source of potash and, therefore, desirable, but it will probably cost more than the product of potash will be worth.

Humus Burning Out.

I would like to know whether or not dry-plowing land, in preparation for sowing oats for hay, injures the soil? I have heard that dry plowing tends to wear out the soil, as the soil is exposed to the sun a long time before harrowing. I have been dry-plowing my land to kill the, weeds, but had a light crop of hay this year.

There is believed to be what is called "a burning out of humus," by long exposure of the soil to the intense heat of our interior districts. It is probable that the reduction of humus is due more to the lack of effort to maintain the supply than to the actual destruction of it by culture methods. Such a little time as might intervene between dry plowing and sowing could not be charged with any appreciable destruction of soil fertility. It is altogether more probable that your hay crop was less from loss of moisture than from loss of other plant food; and it is desirable to harrow a dry plowing, not so much to save the soil from the action of the atmosphere, as to conserve the moisture, which, as you know, will rise from below and will rapidly be evaporated from the undisturbed bases of your furrows. Therefore, we should harrow a dry plowing as soon as practicable, but with particular reference to the moisture supply rather than to other forms of fertility.

Straw for Humus.

Do you consider straw good to plow under for humus, and which kind, wheat, oat, or barley straw, is best?

Straw, by its decay in the soil, produces humus and, therefore acts in the same way just as does the decay of other forms of vegetation. As, however, straw is less easily decomposed than fresh vegetation, it is less valuable and may be troublesome by acquiring a greater amount of moisture by interfering with cultivation or by tending to dry out the soil to the injury of other plants. If the soil is heavy and moisture abundant, straw may be desirable, while in the case of a light soil and scant moisture, may be injurious. There is no particular difference in the straw of the different grains from this point of view.

The Best Legume for Cover Crop.

What would you advise to sow as a crop to plow under? When should it be sowed, and when plowed under?

The best crop for green-manuring in any locality is the one which will make the best growth when surplus moisture is available for it, and when its growth can be undertaken with least interference with irrigation, cultivation and other orchard operation. Generally in California, such a crop can be most conveniently grown during the rainy season, but in some parts of the State where irrigation water is available, a summer growth can be procured with very satisfactory results; so that we are now growing in California both wintergrowing legumes, like field peas, vetches, burr clover, etc., which are hardy enough to grow in spite of the light frosts which may prevail, and are also growing summer legumes which thrive under high temperature, like cowpeas and other members of the bean family, and for which water can be spared without injury to the fruit trees which share the application of the land with them. The plants which are worth trying are burr clover, common or Oregon vetch, Canadian field pea, and the common California or Niles pea. Whichever one of these makes the best winter growth so that it can be plowed under early in the spring, say in February or March, while there is still plenty of moisture in the soil for its decay, without robbing the trees or rendering the soil difficult of summer cultivation, is the plant for you to use largely. All these plants should be sown in California valleys and foothills, as soon as there is moisture enough from rainfall to warrant you in believing they will catch and continue to grow. If the land is light they can be put in with a cultivator and plowed under deeply in the spring, as stated. If the land is heavy, probably a shallow plowing would be better to begin with.

Cowpeas for Cover Crop.

I planted cowpeas between peach trees which I have kept irrigated; when should they be plowed under?

Cowpeas will be killed by frost in most places and should, therefore, be plowed in this fall whenever you have a large growth of green stuff and the ground gets moist enough so that the trees will not be endangered by drying out of the soil, which is likely to occur after plowing in coarse material, unless the soil is kept moist by rain or otherwise.

Garden Peas for Green Manure.

Would it be possible to plant the Yorkshire Hero pea in on orange grove as late as December 25 and get a crop from the peas? Would this pea add much to the fertility of the soil?

You can sow any garden peas as late as December 25, if the ground is in good condition and the temperature not too low. They are grown as a winter crop except when the ground freezes. You would not get as much good for the grove by growing these peas for the market as you would by plowing the whole growth under green, but you certainly will get advantage from the decomposition of the pea straw and of the root growth of the plant.

Grass for Green Manuring.

I wish to sow this fall some green grass to be plowed in next spring
to
improve the soil of part of my land. I read for that purpose a bulle-
tin
I had from the government, but the conditions are so different here
in
California that I am very much puzzled which kind to select.

There is no grass which grows quickly enough to be worth seeding in the fall for spring plowing. It is a good deal better to use a
grain, either barley or rye, for the seed is cheap, the growth quick
and you can get a good deal of green stuff to plow under. Legumes
are, of course, better because of their ability to absorb atmospheric
nitrogen, but any plant which makes a large green growth is good,
and it is better to have a heavy weight of wild vegetation than to
have a light growth of an introduced legume.

Manure with a Clover Crop.

I have an old apple orchard in which I intend to sow burr clover.
In order to get the clover to grow I know that I shall have to use
fertilizer of some kind and this is what I want your advice about.

If you can get it, use stable manure at the time of sowing the clover seed. Stable manure alone will restore the humus and overcome
the rebellious behavior of the soil. Possibly you cannot secure suffi-

cient quantities of it. In that case a little with the burr clover seed will give the plant a good start, or use a complete fertilizer to secure the growth of a legume in the freest and quickest way.

Fenugreek as a Cover Crop.

Fenugreek has been recommended to be as a nitrogen-gathering plant, but I cannot find information as to the amount of nitrogen it gathers in its roots and tops, nor the amount of crop per acre.

Fenugreek is a good nitrogen gatherer and is desirable for green manuring wherever you can get a good growth of the plant. You can count it worth as much as peas, vetches, etc., if you can get as much growth of the plant. It is most largely used in the lemon district near Santa Paula. The best way to proceed would be to try a small area of all the nitrogen gathering plants of which you can get the seed easily, and determine by your own observation which makes the best growth under your conditions.

Improvement of Cementing Soils.

I would like some advice in handling the "cementy" gravel soil. Manure is beneficial in loosening up the soil, but there is not enough available. Would the Canadian field pea make a satisfactory growth here if sown as soon as the rains begin? I would try to grow either peas or vetch and plow under in February or March and then set trees or vines on the land.

The way to mellow your soil is certainly to use stable manure or to plow under green stuff, as you propose. This increases the humus which the soil needs and imparts all the desirable characters and qualities which humus carries. You ought to get a good growth of

Canadian field peas or common California field peas or the common Oregon vetch by sowing in the fall, as soon as the ground can be moistened by rain or irrigation, and, if the season is favorable, secure enough growth for plowing under in February to make it worth while. Be careful, however, not to defer planting trees and vines too late in order to let the green stuff grow, because this would hazard the success of your planting by the reduction of the moisture supply during the following summer by the amount which might be required to keep the covered-in stuff decaying, plus loss of moisture from the fact that the covered stuff prevented you from getting thorough surface cultivation during the dry season. For these reasons one is to be careful about planting on covered-in stuff which has not had a chance to decay. This consideration, of course, becomes negligible if you have water for summer irrigation, but if you expect to get the growth of your trees and vines with the rainfall of the previous winter, be careful not to waste it in either of the ways which have been indicated, and above all, do not plant trees and vines too late. Theoretically, your position is perfect. The application of it, however, requires some care and judgment. Rather than plant too late, you had better grow the green stuff the winter after the trees have been planted.

Needs Organic Matter.

I have what I believe to be decomposed sandstone. Many rocks are still projecting out of land which I blast and break up. The soil works freely when moist or wet, but when dry it takes a pick-axe to dig it up; a plow won't touch it. Among my young fruit trees I tried to grow peas, beans, carrots and beets, and although I freely irrigated them during the summer and fall, and although I planted at different times, my peas and beans have been a total failure, and the beets, carrots and onions nearly so. For years the land has grown nothing but weeds.

Your soil needs organic matter which would make it more easy of cultivation, more retentive of moisture, and in every way better suited to the growth of plants. Liberal applications of stable manure would produce best effects. No commercial fertilizer would begin to be so desirable. If you can dig into the soil large amounts of weeds or other vegetable waste material, you would be proceeding along the same line, but stable manure is better on account of its greater fertilizing content. You ought to be thankful that the soil has spunk enough to grow weeds. The Immanent Creator is still doing the best he can to help you out; take a hand yourself on the same line.

Two Legumes in a Year.

I have land on which I wish to plant to fruits, and I wish to build up the soil all I can, by planting cover crops and plowing under. What would be the best to plant this fall, to be plowed under next spring, and to plant again next spring to plow under in the fall? I will not be able to plant any trees before next fall or the following spring.

Get in vetches as soon as the ground is in shape in the fall. Plow them under early in the spring and close the covering and compact the green stuff by running a straight disk over the ground after plowing. This will help decay and save moisture. Follow with cow peas as soon as you are out of the frost, disking in the seed so as not to disturb the stuff previously covered in. Do not wait to put under the winter growth until it is safe to put on the cowpeas, for, if you do, you will lose so much moisture that the cowpeas will not amount to much.

Handling Orchard Soil.

We average about 35 inches of rainfall. With this heavy rainfall, is there any advantage to be gained by early plowing and clean cultivation right through the winter? Would such plowing and cultivation result in any serious loss of plant food? Would you advise an early or late application of nitrogen, such as nitrate or guano? If there is any loss from an early application, can it be determined by any means?

The old policy of clean winter cultivation has been largely abandoned. Nearly everyone is trying to grow something green during the rainy season to plow under toward the end of it. Even those who do not sow legumes for this purpose are plowing under as good a weed cover as they can get. This improves the soil both in plant food and in friability, which promotes summer pulverization and saves moisture from summer evaporation. Much less early plowing is done than formerly unless it be shallow to get in the seed for the cover crop; the deeper plowing being done to put it under. Guano can be applied earlier in the winter than nitrate, which can be turned in with the cover crop, while the former may be sown with the seed to promote the winter growth. Whether you are losing your nitrate or not the chemist might determine for you by before-and-after analyses. If you are a good observer you may detect loss by absence of the effects you desire to secure.

Soaking Seeds.

Do you think it a good practice to soak seeds before planting?

It is more desirable with some seeds than others and when the ground is rather dry or the sowing time rather late, than when sowing in moister ground or earlier in the rainy season, when heavy rains are to be expected. Soaking is simply a way to be sure that the seed covering has ample moisture for softening and the kernel has what it requires for awakening it germ and meeting its needs. The soil may not always have enough to spare for these purposes and germination may be delayed or started and arrested. Ordinarily seeds can be helped by soaking a few hours in water at ordinary temperatures. Some very hard seeds like those of acacia trees, etc., are helped by hot water - even near the boiling point.

Irrigating Palms.

My palms are quite small, but they do not seem to grow; they seem to be drying up.

The growth of palms is proportional to the amount of soil moisture available, providing it is not in excess and not too alkaline. Some palms are quite drouth-resisting, but it is a mistake to think of a palm as a desert plant and try to make a desert for it. A young palm, especially, needs regular and ample water supply until it gets well established. Your plants may be drying up, or they may have had too much frost or too much alkali. If they are not too far gone, they will come out later if you give them regular moisture and cultivation.

Water from Wells or Streams.

One of our neighbors insists that water from a well is, in the long run, very hard on the land, and that irrigation water is much to be preferred.

There is no characteristic and permanent difference between waters from wells and waters from streams so far as irrigation is concerned. The character depends upon the sources from which both are derived. Some wells may carry too much mineral matter in the form of salt, alkali, etc., and some stream waters sometimes carry considerable alkali. For this reason some wells may be better than streams and some streams better than wells. There is no general rule

in the matter. Your neighbor may be right as applied to your location, and may know from his experience that the well water carries
too much undesirable material. That could only be determined by
analysis, and the analysis must be made when the water is rather
low, because during the rainy season, or soon after it, the water may
have less mineral impurity than later in the season when it may be
more concentrated.

Shall He Irrigate or Cultivate?

Our soil is of an excellent quality, and I feel if the moisture were
properly conserved by suitable methods it could be made to produce fruits or some other very much more profitable than from hay
and grain crops.

Whether you can grow deciduous fruits successfully without irrigation depends not only upon how well you conserve the moisture
by cultivation, but also whether the total rainfall conveys water
enough, even if as much as possible of it is conserved. Again, you
might find that thorough cultivation will give you satisfactory y-
oung trees, but would not conserve moisture enough for the same
trees when they come into bearing. This proposition should be studied locally. If you can find trees in the vicinity which do give satisfactory fruit under the rainfall, you would have a practical demonstration which would be more trustworthy than any forecast
which could be prepared upon theoretical grounds.

Condensation for Irrigation.

If a circular funnel of waterproofed building paper, or some better cheap device, were fastened about the base of the tree in such a

manner as to catch and concentrate most of the drippings from the leaves, and that water made to run down through a tube leading a suitable depth into the earth, it seems to me that the number of foggy nights that occur in many localities during the season might thus supply ample water for a tree's needs.

The probability is that water would not be secured in sufficient quantities to serve any notable irrigation purposes, or if the fogs were so thick as to yield water enough, the sunshine would be too scant for the success of the plant. Put your idea to the test and see how much water you could get from a tree of definite leaf area, which could be readily estimated.

Winter Irrigation.

Last May I irrigated my prune trees for the first time, again during the first two weeks of last December. If no rain should come within the next two weeks, would you advise me to irrigate then? Should I plow before irrigating, or should irrigation be done before the buds swell?

Unless your ground is deeply wet down by the rains which are now coming, irrigate it once, and do not plow before irrigating. The point is to get as much water into the ground and as much grass growth on top as you can before the spring plowing. Never mind about the swelling of the buds. The trees will not be affected injuriously by getting a good supply of winter water into the soil. There might be some danger with trees which bloom late in the spring, like citrus trees or olives, because by that time the ground has become warm and the roots very active. At the blooming time of deciduous trees less danger would threaten, because there is less difference between the temperature of the ground and the water which you were then applying from a running stream. If you irrigated in furrows and, therefore, did not collect the water in mass, its temperature would rise by contact with air, which would be another reason for not apprehending trouble from it.

How Much Water for Oranges?

How much water would you consider absolutely necessary to carry to full-bearing citrus trees an clay loam-that is, how many acres to a miner's inch, figuring nine gallons per minute to the inch?

It would, of course, depend upon the age of the trees, as old bearing trees may require twice as much as young trees. We would estimate for bearing trees, on such retentive soil, 30-acre inches per year applied in the way best for the soil.

Damping-off.

My orange seed-bed stack has "damp-off." Same say "too much water;" "not enough water;" "put on lime;" etc. I use a medium amount of water and more of my stack is affected than that of any other grower. One man has kept his well soaked since planting, and only about six plants were affected. Another has used but little water, keeping them very dry; he has lost none.

Damping-off is due to a fungus which attacks the tender growth when there is too much surface moisture. It may be produced by rather a small amount of water, providing the soil is heavy and the water is not rapidly absorbed and distributed. On the other hand, a lighter soil taking water more easily may grow plants without damping-off, even though a great deal more water has been used than on the heavier soil. Too much shade, which prevents the sun from drying the surface soil, is also likely to produce damping-off, therefore, one has to provide just the right amount of shade and the right amount of ventilation through circulation of the air, etc. The use of sand on the surface of a heavier soil may save plants from damping-off, because the sand passes the water quickly and dries, while a heavier surface soil would remain soggy. Lime may be of advantage if not used in too great quantities because it disintegrates the surface of the soil and helps to produce a dryness which is desirable. Keeping the surface dry enough and yet providing the seedlings with moisture for a free and satisfactory growth is a matter which must be determined by experience and good judgment.

Irrigated or Non-Irrigated Trees.

Is there any difference between the same kind of fruit trees grown without irrigation and with it?

It does not make a particle of difference, if the trees are grown well and matured well. Overirrigated trees or trees growing on land naturally moist may be equally bad. Excessively large trees and stunted trees are both bad; with irrigation you may be more likely to get the first kind; without it you are more likely to get the latter. There is, however, a difference between a stunted tree and a well-grown small tree, and as a rule medium-sized trees are most desirable than overgrown trees. The mere fact of irrigation does not make either good trees or bad trees: it is the man at the ditch.

Too Little Rather Than Too Much Water.

Looking through an orchard of 18-year-old prune trees on river-bottom land, I found a number of the trees had died. A well bored in the orchard strikes water at about 15 feet. I find no apparent reason far the death of these trees unless it is that the tap roots reach this body of water and are injuriously affected thereby.

We do not believe that water at 15 feet depth could possibly kill a prune tree. It is more likely that owing to spotted condition of the soil, gravel should occur in different places, and with gravel three or four feet below the surface a tree might actually die although there was plenty of water at a depth of 15 feet. There is more danger that the trees died from lack of water than from an oversupply of it, and it is quite likely also that you could pump and irrigate to advantage large trees which did not seem to be up to the standard of the whole place, as manifested by lack of bearing, smallness of leaves, which would be apt to turn yellow too early in the season.

Possibly Too Much Water.

My trees are four years old and are as follows: Peach, fig, loquat, apple, apricot and plum. Last year they had plenty of blossoms, but I got no fruit. I always watered them twice a week in summer.

You are watering your trees too much; stimulating their growth too much, and this, while a tree is young, is apt to postpone its fruit bearing. Give the soil a good soaking about once a mouth, unless you are situated in a sandy or gravelly soil, in which more frequent applications may be necessary.

Too Little Water After Dynamiting.

In planting almonds on a dry hard soil I dynamited the holes and ran about 200 gallons of water into each hole before planting. About 95 per cent of the trees started growth, but seem now to be in a somewhat dormant state, the leaves of some being slightly wilted. All the trees were watered since planting. I have been told I made a mistake by throwing water in the dynamited holes. When the holes were watered the ground was very dry and the water disappeared in a few minutes.

You have used too little water rather than too much. Dry soil of fine texture can suck up an awful lot of moisture, which can be drawn off so far, or so widely distributed, that there will not be enough for the immediate vicinity of the roots. The dynamiting tended to deep drying and necessitated much more irrigation.

Irrigating Young Trees.

We have just put out 50 acres to walnuts. The party who put them out wants me to have some boxes or troughs made 15 inches long with a 3-inch opening, and put in on the slant so as to have the water hit the roots.

Many such arrangements of boxes, perforated cans, pieces of tile, etc., have been proposed during the last fifty years in California for accomplishing the purposes which are mentioned in your letter, and all such devices have been abandoned as undesirable. They may bring the water to bear upon a lower level as intended, but the free access of air and the fact that, with their use, proper stirring of the soil is neglected renders them undesirable. The best way to water young trees singly is to make a trench around tree, but not allowing the water to touch the bark, applying the water and then thoroughly hoe when the surface soil comes into proper condition. Young trees treated in this way, with the surface always in good condition, do not require much water. The amount depends, of course, upon whether the soil is naturally porous or retentive.

Underground Irrigation.

How extensively used and with what results is the underground tile system for irrigation used, and what especial character of soil is it best suited for?

Not extensively at all; in fact, if there is an acre of it which has been for three years in continuous and successful operation, it has escaped us. After forty years of trial of different systems, none has demonstrated value enough to warrant its use. Theoretically, they are excellent; in practice they are defective. Surface application in different ways, according to the nature of the soil, accompanied with thorough cultivation, is the only thing that at the present time promises satisfactory results, except that where the land suits it, irrigation by underflow from ditches on higher elevations is being successfully used on small areas in the foothills. For gardens the most promising arrangement seems to be a laying of drain tiles rather near the surface, which shall be taken up each year, cleaned of silt and plant roots, and relaid along the rows before planting; but this calls for too much labor, except perhaps for amateur gardeners. The kind of soil best suited to such a system is a medium loam which will distribute water sufficiently to avoid saturation and air-exclusion. Both a heavy soil which does this, and a coarse sandy loam which takes water down out of reach of shallow-rooting plants too rapidly and lacks capillarity to draw it up again, are ill adapted to underground distribution.

Irrigation of Potatoes.

Will you kindly tell me when is the proper time to irrigate potatoes, before they bloom or after they bloom, and do they require much water?

It should seldom be necessary to irrigate potatoes after the bloom appears. Potatoes do not need much water, and there is danger of giving them too much. It is absolutely essential to see that there is no check in the growth of the plant, for once the growth is at all checked by drought, and irrigation is done, a new lot of potatoes start and new and old growth of tubers are worthless. Give what irrigation is needed and make cultivation do the rest. The secret of success is keeping the soil continually at the right moisture, so that

the first growth of the plant may continue regularly until the tubers
are brought to maturity.

Irrigated or Non-Irrigated Apples.

Where soil and climatic conditions are favorable to the raising of
apples, what effect has irrigation an them?

The commercial product of California apples is chiefly made upon deep soils in districts of ample rainfall so that the fruit can be
perfected and the trees maintained in thrift by thorough cultivation
and without irrigation. In the foothill and mountain regions, however, apple trees are irrigated and first-class fruit produced by the
process. There is no particular virtue in the absence of irrigation nor
in the presence of it. All that the tree requires is that the moisture
supply should be adequate and timely. There are undoubtedly
many apple orchards grown without irrigation where a little water
during the latter part of the summer would be a great advantage for
the perfection of winter varieties.

Irrigating Walnuts-Checks or Furrows.

Which is the best method to irrigate a tract of 25 acres of sandy
sediment sail, nearly level, preparatory to planting walnuts?

By all means use the furrow system of irrigation unless your land
should be so light that the water would sink in the furrows and
distribution would be very unequal without covering the whole
surface as is done by filling checks. When the land cannot be covered well by the furrow system, checking is resorted to, but not
otherwise.

Summer and Fall Irrigation.

Is it desirable to irrigate peach trees in the fall after the crop is gathered?

The popularity of autumn irrigation for peaches in the San Joaquin valley is based upon the experience of the last few years where trees that have been allowed to become dormant too early in the season and have been weakened by a long period of soil-drought during the autumn, have cast their blossoms or manifested other indications of weakness during the following year. It is thoroughly rational to apply irrigation to hold the leaves and secure their service in the strengthening of bloom buds for the following year by irrigation. Such irrigation should be applied immediately after the fruit is gathered or even before that, if the yellowing of the leaves indicates lack of strength in the tree and the frequency and amount of irrigation during the autumn depends upon whether the soil will hold moisture enough to carry the tree to its proper period of dormancy. This may be determined by the aspect of the trees and by digging down two or three feet to see whether the soil carries moisture which is likely to be sufficient until the coming of the rains. Whether late irrigation will be necessary is also determinable by the character of the soil; on close retentive soil it may not be necessary, while on loose, sandy or gravelly soil it may be essential to the life of the tree. One has to settle all these matters by judgment and not by recipe.

Fertilizers in Irrigation Water.

Do you recommend putting fertilizers in irrigating water? I am about to water the orchard and am thinking of putting some nitrate in the water.

You can distribute any soluble fertilizer by dissolving it in irrigation water, but few have ever done it because of the difficulties of getting equal strength in running water. It is much easier to distribute on land before irrigation.

Irrigating Alfalfa on Heavy Soils.

How does alfalfa succeed on adobe and soils slightly modified from it?
Does irrigation work well an adobe planted to alfalfa?

If you get the irrigation adjusted so that the soil shall not be water-logged and so that the water does not stand on the surface when the sun is hot, you can get plenty of good alfalfa on a heavy soil. Irrigation on adobe soils must be done more frequently and a less amount at each application to guard against the dangers named above.

How Much Water for Crops?

Same of my land is heavy but the most of it is light soil. I want alfalfa mostly, same potatoes and grain, and later oranges, olives and other fruit. How much water in inches or acre feet is required per acre per year far the irrigation of it?

The amount of water required to grow different crops depends upon the crop itself, upon the time of the year in which it grows, the character of the soil, etc. There is no such thing as stating how much water would be used for all crops on all soils, and at all times of the year. The range would be from, say, ten acre inches for irrigation of deciduous fruits, which need moisture supplementary to rainfall;

twice or thrice as much for citrus fruit trees; four or five times as much for alfalfa where a full number of cuttings are required. These are, of course, only rough estimates which would have to be modified according to local rainfall and soil character. Water should be applied frequently enough to keep the lower soil amply moist. A color of moisture is not enough and a muddy condition results from too much water. One has to learn to judge when there is moisture enough, and a good test of this to take up a handful of soil, squeeze it and open the hand. If the ball retains its shape it is probably moist enough. If it has a tendency to crack upon opening the hand, it is too dry. This test, of course, is somewhat affected by the character of the soil, but one has to form the best judgment possible how far allowance has to be made for that.

Sewage Irrigation.

What is the usefulness or harmfulness of the outflow from septic tanks for use an fruits and vegetables?

There is no question as to the suitability of the affluent from a septic tank for irrigation purposes. Waste waters are sometimes injurious when they are loaded with antiseptics, but the septic tank will not work unless it has a chance for free fermentation in the absence of antiseptics, therefore, this objection against waste water does not hold with the out-flow from septic tanks. It has the advantage over straight sewage irrigation because fermentation in the septic tank is believed to free the water from many dangerous germs, though not all of them.

Creamery Wastes for Irrigation.

Will the waste water from a creamery, pumped into a ditch and used for irrigating sandy loam orchard land, or nursery stack, in any way be injurious to the land or the trees?

It will depend upon the amounts of salt and alkaline washing materials which it carries. This would be governed, of course, by the amount of fresh water used for dilution in the irrigation ditch. There are two ways to determine the question. One would be to make an analysis of a sample of the water taken when it contains the largest amount of these materials after the dilution with ditch water. Another way would be to plant some corn, squashes, barley and other plants, so that they would be freely irrigated by the water during one growing season. This would be rather better than an analysis, because everybody could see whether the plants grew well or not, and would be apt to be better convinced by what they see than by an opinion which a chemist might give on the basis of an analysis. The use of this water on a sandy loam would obviously be less injurious than upon a heavy retentive soil.

House Waste Water.

Is it feasible to use wash water, etc., for watering fruit trees and vegetables?

Kitchen sink water is not desirable because of its great content of grease, but wash-tub and bathtub water are good. Strong soapsuds should be mixed with considerable rinsing water to escape excessive content of alkali. Run the water in hoe-ditches, along the rows of vegetables, hoeing thoroughly as soon as the land hoes well, changing the runs of water so that the soil does not become compacted but is kept friable and lively.

Draining a Wet Spot.

I have a spot of about an acre that in a wet winter becomes very miry and as a rule is wet up to July. Can I put in a ditch two and one-half feet deep and fill in with small stones for a foot or a foot and a half, until I can afford to buy tiles?

Drains made of small stones are often quickly filled with soil and stop running. However, it will work for a time, and such drains were formerly largely employed in Eastern situations when cash was scant and stones abundant. Dig the ditch bottom to a depth of not less than 3 or 3 1/2 feet, then put in the stones deep enough not to be interfered with by plowing. If you have flat stones you can make quite a water-way with them and fill in with small stones above it.

Part V. Live Stock and Dairy

Legal Milk House.

What is a legal milk house in California?

The State dairy law says little concerning the construction or equipment of the milk house. It says that the house, or room, shall be properly screened to exclude flies and insects, and is to be used for the purpose of cooling, mixing, canning and keeping the milk. The milk room shall not be used for any other purpose than milk handling and storing, and must be 100 feet or more distant from hogpen, horse stable, cesspool or similar accumulation of filth, and must be over 50 feet from cow stalls or places where milking is done. In regard to the size of the milk room and equipment, nothing is said provided it is large enough for the milk to be handled conveniently. Concrete milk houses, however, had best have smooth-finished floors and walls. The interior of the milk house is also to be whitewashed once in two years or oftener. If milk from the dairy is to go to a city, the requirements will be more severe than provided in the State law, and must conform to the ordinances of the city to which the milk is to be sent.

Cure for a Self-Milker.

What shall I do for a young cow that milks herself?

Fit a harness consisting of two light side slats and a girth and neck strap in such a way that the cow cannot reach her udder. Unless she is particularly valuable for milk, it will save you a lot of worry to fix her up for beef.

Strong Milk.

How can I overcome strong milk in a three-quarter Jersey cow? I had been feeding alfalfa hay with two quarts alfalfa meal and one quart middlings twice a day. Thinking the strong milk came from the feed I changed to oat hay and alfalfa with a soft feed of bran and middlings.

There is nothing in either ration that could cause strong milk, nor will a change of feed likely benefit the trouble. If the cow is in good physical condition the trouble probably comes from the entrance of bacteria during or after milking. Thoroughly clean up around the milking stable, followed by a disinfection of the premises. Have the flanks, udder and teats of the cow thoroughly cleaned before milking and scald all utensils used for the milk. Harmful bacteria may have gotten well established on the premises and the entrance of a few is enough to seriously affect the flavor of the milk. Once the trouble is checked it can be kept down with the usual sanitary methods.

Separator as Milk Purifier.

I have a neighbor who contends that a cream separator purifies the milk that passes through it. I say that it does not purify the milk. I agree that it does take out some of the heavy particles of dirt and filth, but that it cannot take out what is already in solution with the milk.

The purification naturally cannot be very great, and if milk is produced in unsanitary fashion, running through the separator will do little, if any, good. Nevertheless, the separator does remove more than just the solid particles of dirt. The purifying comes by leaving behind the separator slime, so called, the slimy material left behind after a good deal of milk has been run through. In fact, some creameries separate milk, only to mix milk and cream again, largely for the purpose of removing the impurities found in the slime. In this slime are not only the impurities that fall into the milk, but also some of the fibrous matter that is part of the milk, and this gathers, being pulled out by gravity as are the fat particles, it seems to gather with it a few more bacteria than remain in the milk itself. Material in real solution, as sugar is in solution in water, naturally is practically unaffected by separation. You are, therefore, right to the extent that you cannot produce unsanitary milk and clean it with the separator, but your neighbor is right to the extent that the separator does remove some impurities and is used just for that purpose. There is also in the dairy trade a centrifugal milk clarifier which is constructed in somewhat similar manner to a cream separator, but acts differently on the milk in not interfering with cream rising by gravity when separated cream and milk are mixed after cleaning.

Butter Going White.

I bought some butter and during the warm weather it melted. About 40 or 50 per cent was white, while the balance was yellow and went to the top. When the butter remelted, the yellow portion melted, leaving the white portion retaining its shape. The white portion did not taste like ordinary butter. The butter made from our cows' cream melted at a higher temperature, but did not have a white portion. Why did our butter not act like the creamery butter?

Samples of butter have occasionally been sent to this office that have turned white on the outside, and since the white part has a very disagreeable, tallowy flavor, people think that tallow or oleomargarine has been mixed with it, but we have never been able to find any foreign substance in any of the samples. We have found that some of the best brands of butter will turn white first on the outside and the white color will gradually go deeper if the butter is exposed to a current of air or if left in the sun a short time - F. W. Andreason, State Dairy Bureau.

What Is "Butter-fat?"

I would like to know what "butter-fat" means. I have asked farmers this question and no one seems to know. I suppose all parties dealing with creameries understand what the standard of measure or weight of butter-fat is, but it is my guess that there are thousands of farmers whom, if they were asked this question, would not know. We, of course, know that butter is sold by the pound and cream by the pint, quart or gallon, but what is butter-fat sold by?

Butter-fat is the yellow substance which forms the larger part of butter. Besides, this fat butter is composed of 16 per cent or less of water and small amounts of salt, and other substances of which milk is composed. From 80 to 85 per cent or so of ordinary butter is the fat itself. It is sold by weight. The cream from which butter is made is taken to the creamery and weighed, not measured. A small sample is tested by the so-called Babcock test to determine the exact percentage of fat, and payment mode on this basis. For instance, if 1,00 pounds of cream is one-third butter-fat, the dairyman receives pay for 33 1/3 pounds of this substance. If it is only one-quarter fat, he receives pay for 25 pounds. Ordinary cream varies within these limits, but may be much richer or thinner. Cream after the butterfat is removed is much like skimmed milk, although it has less water in it.

Why Would Not Butter Come?

What is the trouble with cream that you churn on from Monday until

Saturday, then have to give up in despair and turn it out to the hogs?

We warmed it, and we cooled it, and used a dairy thermometer, but nothing would do.

If the cream was in churnable condition otherwise, the probability is that it was too cool when you started churning. It should be about 62° Fahrenheit.

Drying a Persistent Milker.

My cow is to come fresh about the middle of next mouth, and in the last two weeks her milk has changed in some way so that the cream makes very yellow butter and comes to butter nearly as quick as when the cow was fresh. Would it best for her to go entirely dry before coming fresh, or will it be all right if she does not entirely dry up?

If your cow has been able to pick up any special amount of grass since the rains came it might add to the color of the butter. A cow's milk also gets richer toward the end of her lactation period, which may make a richer cream and make the butter come quickly There does not seem to be anything to worry about. The cow would probably do better if she could become entirely dry before calving, but unless you can easily dry her up it would be dangerous to try to force her to do so.

Butter-fat in Sweet and Sour Cream.

The creamery wagon takes our cream every other day. Without ice it is almost impossible to keep the cream sweet during the hot weather. By the time the wagon gets here, several hours after the fourth milking, the cream is quite sour. Does sour cream test lower than sweet cream! Is any butter-fat lost due to evaporation in dry weather?

The test of sour cream will be as accurate as of sweet cream, if properly made, but it is rather more difficult to make; or rather, to get the material into condition to work well. There is no fat lost by evaporation.

Cream That Won't Whip.

When I sell my cream from the separator they say they cannot whip it.
Can you tell me if there is any way that I can make the cream whip?

There appears to be no good reason for blaming the separator for your difficulty with the cream. Possibly the cream may be too thin, as thin cream is sometimes difficult to whip. There is also the possibility that the fat globules in the cream may be rather small, but that will be the fault of the cows, not of the separator. Another reason why the cream may not whip well may be that it is used too quickly. If the milk is all right, the cream not too thin and it is permitted to stand for 12 hours or so there should be no trouble with it. Occasionally when cream is pasteurized it will not whip well. In these cases, or any other that may develop, the application of lime water to the cream at the rate of 1 gallon to 60 will remove the difficulty.

What Is Certified Milk?

What process has milk to go through to be called "certified," and what demand is there for it?

Certified milk is simply milk that is produced and marketed under prescribed sanitary conditions. The dairies are inspected periodically by representatives of some medical society or other organization to see that all regulations are observed, who certify that this is done; hence the name. Milk from other dairies is prohibited by law from being sold under the name "certified milk." Among the requirements in its production are that the cows must be free from tuberculosis and otherwise perfectly healthy, the stable to have a concrete floor which is washed out after each milking, the milkers to have special clothes for milking, etc. The milk is cooled and bottled immediately after milking, and kept at a low temperature until it reaches the consumer, to prevent the entrance of dirt of any kind or the development of the few bacteria that must gain entrance before it is bottled. To produce such milk requires much expensive apparatus and much more labor than to produce ordinary milk, and as a result it sells for a much higher price, both to distributor and consumer, so that the market for it is rather limited.

Jersey Shorthorn Cross.

If I cross Registered Shorthorns with a Jersey bull, what dairying value will the progeny have?

This makes an excellent cross. Even beef-strain Shorthorns have lots of milking power if it is developed and the Jersey cross will bring it out in the progeny. The cows have excellent milking qualities and give very rich milk. They also have a big frame and fine constitution. About the finest cows in Humboldt county were of this cross although Jersey bulls have been used so long that the Shorthorn blood is almost eliminated. The first "improved" cattle in California and the first cross made for dairy purposes was Jersey bulls upon grade Shorthorn cows. Later the Holstein Friesians became popular and they and their grades are now most abundant.

A Free Martin.

I have a Jersey cow who has just had twin calves, a heifer and a bull. The heifer was born about five minutes before the bull and seems to be the stronger. My neighbors tell me to fatten both for the butcher, for they say the heifer will be barren. The mother is a young cow, as this is her second calf. Kindly inform if this is one of nature's laws or if there is a possibility of the heifer turning out all right?

The probability is that it will be better to veal the heifer than to raise her, as most heifer calves twinned with a bull are free martins, or animals of mixed sex and no good for breeding purposes or for profitable milk production. If the bull is a good animal, he probably will be all right, as this twinning does not seem to affect a bull calf, though it does the heifer. It does not always happen that the heifer is worthless for breeding, but the probability is so great that you had better have her killed and be done with it.

What Is a "Grade"?

Does the term "grade" mean an animal whose sire is a thoroughbred and whose dam is a scrub, or just one who is selected from others because of her good points or those of her mother?

Roughly speaking, a grade animal is one having more or less pure-bred blood, but not enough, or otherwise too irregular, for registry under the rules of the association of the breed to which it has affiliation. It does not refer to selection without use of a pure-blood sire at some point in the ancestry, but this is not a distinction of much moment, for it is hard to find animals which have not borrowed something from some cross with pure blood, though remote. The terms high and low grade are sometimes used to signify amount of pure blood recognizable by form and other characters or remembered by owners or their neighbors. Generally speaking, a grade is anything not entitled to registry, though ordinarily it refers to the offspring of a pure-bred sire and a cow of another or of no breed. The offspring of a pure-bred cow and a scrub bull would also be a grade.

Breeding a Young Mare.

I have a beautiful colt 22 months old that will weigh 1200 or 1300 pounds; very compactly built, and has extra health, life and vigor. I want this colt for a broodmare. Would you advise breeding at two or three years old?

Authorities agree at placing the age from two to three years, according to the development of the animal and other circumstances.

"To Breed in the Purple."

What is meant by breeding a sow in the purple? I have seen this statement used many times by breeders who advertised "sows safe in pig bred in the purple."

To be "bred in the purple" means to be of royal or princely parentage. It originally was used in reference to the nobility of Europe, as purple was the insignia of royal blood, due to the fact that purple was the rarest and most costly color and only the rich and noble could buy it. When used in referring to live stock, it signifies that the animal in question has a long line of blooded ancestry.

Cows for Hill Country.

What breed of dairy cows do you think would be preferable to keep for butter, at an altitude of about 1800 feet, in Nevada county - Jerseys, Guernseys or Ayrshires? I do not mean to have them to rustle for their own living, but to feed them well, house and care for them in all weather, particularly in stormy weather.

The best breed for a man is the one he likes best, providing it has been bred for the purposes he desires to attain. All the breeds you mention are suited to the scheme you outline.

Foothill Dairying.

Is there any risk to run in taking cows to an altitude of 2000 from a much lower one?

There is no quarrel between a cow and a mountain. Ever since the settlement of the State cows have been driven directly from the valley up to the mountain meadow pastures, both for butter and for beef-making, in the summer time. The foothill elevation you mention is only a starting to elevations of 6000 feet and more to which cattle are driven every season.

Bad-Tempered Jerseys.

Jersey bulls are apt to become vicious after a time; is it so to the same extent with bulls of the other named breeds?

The Jersey bull is conceded to be crosser and more dangerous than other bulls, but no bull should ever be allowed to have a chance at a man. Never consider a bull gentle and you will be safe with him.

Breeding in Line.

Is it right and proper to breed a pedigreed registered bull to his daughter, who is the offspring of a grade cow? If it is not right, explain why. If it can be done, will the offspring be physically perfect and an improvement, or will it have poorer qualities than its sire and mother? If this inbreeding can be done successfully, how long can it be carried on, or, in other words, how long could one bull be bred back into his own offspring? Can a herd be perfected in this way?

It is right and proper to breed a registered sire to his daughter, who is the offspring of a grade cow. The first cross is all right and the offspring ought to be physically perfect. This is a first step in what we call line breeding, but in line breeding proper, both animals must be pure bloods and registered, having ancestors on both sides which have a long line of good individuals with strong constitutions and true to type. To do this, one must have a perfect ideal in mind. This line breeding is what has developed the breeds today up to the high standard of perfection. Breeding sire to daughter, if followed along these lines, will be all right; at least, it was so in the case of Amos Cruickshank, the great shorthorn breeder. You cannot successfully breed back on the daughter's offspring, but if you use a straight out-cross on the daughter's offspring you can again use this sire on her produce with marked success. In the case of a grade cow and registered sire, there are two things which will make you either lose or win with one cross, and that is regarding the breeding of your sire. If he is just an ordinary-bred fellow it will be a hit-and-miss game, but if he is from a long line of good ancestors on his dam's side, you can very materially improve the, herd, because always keep in mind the female produce from the sire's dam will grow with age toward the sire's dam. So if your first cross from your first sire is all right, use a straight out-cross bull, but be sure he is what he ought to be, and then you can use your old bull back on his heifers. Of course, a man practicing this breeding ought to be a thorough stockman and a first-class judge of live stock. - W. M. Carruthers.

Whitewashes for Stock Buildings.

I desire whitewash recipes which have given durable results on outbuildings.

It is so desirable to make outbuildings neat and clean, and so important to keep trees from sunburning, etc., that a durable whitewash as cheaply and easily made as possible is very important. The following are commended: No. 1 - To half a bucketful of unslaked lime add 2 handfuls of common salt, and soft soap at the rate of 1 pound to 15 gallons of the wash. Slake slowly, stirring all the time. This quantity makes 2 bucketfuls of very adhesive wash, which is not affected by rain. No. 2 - Whitewash requires some kind of grease in it to make it most durable. Any kind of grease, even though it be old and partly spoiled, will answer all right, though tallow is best. The grease imparts to the whitewash an oil property the same as in good paint. Tallow will stay right on the job for years, and the cheapest of it will do. In order to prepare this grease and get it properly incorporated into the white wash, it is necessary to put the grease in a vessel on the stove, and boil it into a part of the whitewash so as to emulsify it and get it into such condition that it can be properly incorporated with the whitewash mixture. No. 3 - For every barrel of fresh lime, add 16 pounds of tallow, 16 pounds of salt and 4 pounds of glue, dissolved. Mix all together and slack; keep covered, and let stand a few days before using. Add water to bring the right consistency to spread readily. For nice inside work strain it. When less than a barrel of lime is used, the quality of the wash does not seem so good. It is better to apply hot, but it does well cold.

Government Whitewash.

What is the government recipe for whitewash?

"Take a half bushel of well-burned, unslaked lime, slake it with boiling water, cover during the process to keep in steam, strain the liquid through a fine sieve or strainer, and add to it 7 pounds of salt, previously dissolved in warm water; 3 pounds of ground rice boiled to a thin paste and stirred in while hot; half a pound of Spanish whiting and 1 pound of glue, previously dissolved by soaking in cold water, and then hanging over in a small pot hung in a larger one filled with water. Add 5 gallons of hot water to the mixture, stir well and let it stand for a few days, covered from dirt. It should be applied hot, for which purpose it can be kept in a portable furnace. A pint of this mixture, if properly applied, will cover a square yard."

Whitewash for Spray Pump.

Can you give a recipe for a durable whitewash which can be prepared simply and in large quantities? The whitewash will be applied with a spray pump.

To 25 pounds of lime, whole, slacking with 6 gallons of water, add 6 pounds of common salt and 1 1/2 pounds of brown sugar. Stir and mix well and allow to cool. When cool stir in 1 ounce of ultramarine blue. Then add 2 gallons of water, and sprinkle and stir in 2 pounds of Portland cement. If two coats are to be applied, add 1 more gallon of water. Strain for work on smooth surface.

Buttermilk Paint

How is paint made with buttermilk for farm buildings?

One gallon buttermilk, 3 pounds of Portland cement, and sufficient coloring matter to give the desired shade. Apply as soon as made, and stir a great deal while being applied. It is said to dry in about 6 hours and to be a good preservative for fences, barns and other outbuildings.

Trespassing Live Stock.

Is there a fence law in this State? In other words, do I have to fence against my neighbors' stock, or does the law require him to care for his stock and keep it off my property?

The old "no-fence law" which was enacted during the troubles between wheat growers and stock rangers has been put out of commission by more recent legislation. The trespassing live stock is liable for damage, but just how to proceed to protect yourself you should learn from a local lawyer who knows statutes and your county ordinances also.

Rat-Proof Granary.

How can I make a rat-proof granary for alfalfa meal and barley?

Omit all boarding of the sides below the floor level and place a heavy inverted pan, milk pan, between the top of each of the supporting posts and the floor beams. Care should be taken that the diagonal bracing of the underpinning or posts does not allow a rat to secure a foot hold near enough the floor to permit of gnawing through.

Concrete Stable Floor.

Is a concrete floor good for a horse stable?

Concrete floors are satisfactorily used for horse stables, provided the floor is ribbed or otherwise roughened in a way to reduce the danger of slipping. Some stablemen have stall floors made that way. Some use a wooden grating over the concrete in places where the horses have to stand for any length of time. Others soften the standing by free use of bedding.

Silo-Heating Not Dangerous.

Is there any danger of a barn burning from spontaneous combustion due to a silo being built in the barn?

There is no danger of the silo overheating and setting fire to a barn. When the ensilage is curing, it often gets warm, but never anywhere near the point of combustion.

To Make Shingles Durable.

What is the best material with which to coat the shingles on my barn roof?

The best coating is a wood preservative, the principal ingredient of which is creosote. There are several reliable brands of preservatives and stains that may be had at a cost of about half that of paint. We must remark also the natural durability of redwood shingles in this climate if the roof has a good pitch. We reshingled our house roof after 20 years of use and found the shingles so sound that we turned them and shingled the sides and roof of a shed with them where they promise to be good for another score of years.

Best Breed of Hogs.

What is the best breed of hogs for pen feeding, shutting them up in small pens from the time they are little pigs and feeding them mostly on skim milk and slops?

There is no best breed. It is a matter of personal preference. Any of the breeds are all right to pen up and feed. The principal thing is to see that the hogs are all pure bred and have not been crossed too often to cause deterioration. Choose one breed of hogs and keep them as pure as possible and you will have no trouble in raising them. All the breeds are good; but some are fancied more than others. Dark-colored hogs are preferred in California because less liable to sunburn.

Part VI. Feeding Farm Animals

Feed for Plow-Horses.

While doing heavy plowing, how many pounds of rolled barley per day should I feed to keep 1300-pound horses in good condition? If I feed part oat hay and part alfalfa hay, together with rolled barley, what ration would be ample?

A ration used by the California Experiment Station was 12 pounds of alfalfa hay, 11 pounds of wheat hay and 7 pounds of crushed barley for 1000 pounds of horse at hard work. The larger the horse the less food for the amount of work he does in proportion to his size, so multiplying these figures by 1.2 would bring a person somewhere near the ration for a 1300-pound horse, and an approximation is as close as one can come to any general ration. Probably more alfalfa and less of the other feeds could well be given, since many farmers are succeeding in feeding alfalfa exclusively.

Vetch for Horses.

Does vetch make good feed for horses? Will vetch produce a heavier crop than grain? When is the best time to sow vetch for hay, and what is the best variety?

Vetch makes excellent stock feed whether used as hay or as pasturage. Vetch falls to the ground so badly that it is very difficult to cut hay from it unless some grain is planted to hold it up. Oats make an excellent hold-up crop and is more generally used. A half a bushel of vetch seed is mixed with a bushel of oats and this is enough to plant an acre. Some growers, however, prefer a bushel of vetch as that makes the stand much heavier.

Sorghum Feeding.

Can I allow milk cows to pasture on growing Kaffir and Egyptian corn during the summer? Which one is the best for pasture and milk?

There is no difference between Kaffir corn and Egyptian corn so far as feeding goes. They are both sorghums. There is a danger in pasturing on young sorghums, because stock is often killed from overeating it, and they are quite apt to do this when they come upon it from dry feed. If you cut and wilt the young sorghum, or if it is fed sparingly with hay, etc., it becomes innocent of injury. After the sorghum has obtained considerable growth, it also loses its dangerous character.

Salting Hay.

What kind of salt is used for salting hay, how much to use and how to apply it?

Any good commercial salt such as is used for pork or beef packing is satisfactory for salting hay. A good handful to the ton, scattering it as the hay is stocked is as good a formula as can be had.

Stover.

What is stover? How is it cut and handled?

Stover is corn fodder after the ears are taken off. The best time to cut the corn for stover is immediately after the kernel becomes dented and the leaves or blades commence to dry. Immediately after the ears are taken off, the stalks should be cut and stacked. The size of the shock depends upon the climate. If it is a foggly climate and stalks are green, it is better to make a smaller shock, but in the interior valley where the weather is warm it is best to make large shocks, so that the stacks will not dry up very rapidly.

Feed for Cows.

What shall I feed cows when they are fresh and when they are dry!

When they commence to freshen, give some green feed, such as alfalfa or corn; if possible, also give, say, two or three pounds of barley or bran, and gradually increase this for two or three weeks until six or seven pounds of bran or barley is being fed. Also give a small amount of hay. Bran may be rather expensive feeding and a substitute is being used. Take four parts of barley to one of bran and mix. With barley at its low price, this makes rather inexpensive feeding. Another substitute is to take the chopped alfalfa hay and barley. These are mixed thoroughly together and moistened. After the cow freshens and gives her full flow of milk, let her eat all the alfalfa hay she wants. A good ration is about 15 to 20 pounds of hay, 6 or 7 pounds of barley or bran and about 10 pounds of roots such as beets or mangels. When the cow is dry, pasture is the best food, supplemented with some green food.

Sorghum Silage.

Will Egyptian corn make good ensilage and at what time should it be cut to make the best feed for dairy cows?

Sorghum makes good silage. It must be cut while surely juicy enough, for it is a little more apt to dry out than Indian corn.

Barley for Hay Feeding.

Should the barley for hog feeding be rolled, ground or fed whole, dry or wet? Also, how much should be fed and how often to get best results?

To obtain the best results, the barley should be ground into a meal (not too fine) and have the hulls screened or floated out. This is best fed when made into a thick slop. Some good feeders believe in letting it stand until fermentation sets up, that is, gets a little sour. We prefer a sweet to a sour feed. However, hogs will do well on either, provided there is no change from sour to sweet. The change is the bad part. Hogs should be fed just the amount that they will clean up well, and no more. A hog should always be ready for his feed at feeding time. We would not feed oftener than twice a day: night and morning. - Chas. Goodman.

Sugar Beets and Silage.

Will sugar beets keep in a silo and how sugar beets rank as a hog feed?

Sugar beets would probably keep all right if stored in a silo just as they might if kept in any other receptacle, but it is not necessary to store beets for stock-feeding in this State. They can be taken from the field, or from piles made under open sheds in which the beets may be put because more convenient for feeding than to take them from the field in the rainy season. Beets put whole into a silo would not make silage. For that purpose they would need to be reduced to a pulp, but there is no object in going to the expense of that operation where beets will keep so well in their natural condition and where there is no hard freezing to injure them. Beet pulp silage is made from beets which are put through a pulping process for the purpose of extraction of the sugar and, therefore, best pulp silage is only made in connection with beet-sugar factories and is a by-product thereof which is proving of large value for feeding purposes.

Feeding Value of Spelt.

What is the food value of spelt? It is a Russian variety of wheat, and yet, I am informed, it has about the same value as a stock food that barley has.

We have no analysis of spelt at hand. It is presumably like that of barley, as you suggest, because the spelt has an adhering chaff as barley has. This fact makes it better for feeding than wheat, not in nutritive content, but because the chaff tends to distribute the starchy material, making it more easily digestible; just as barley and oats are better than ordinary wheat for stock feeding.

Concentrates and Corn Stalks.

Is it necessary to feed mulch cows any hay or concentrated feed in addition to green corn stalks?

It is necessary. Green corn is an excellent thing for milch cows, but it is a very unbalanced ration and needs alfalfa or something else to balance it up. Green corn, for example, contains only about one per cent of digestible protein and 11.5 per cent of digestible carbohydrates and 0.4 per cent fat, or a nutritive ratio of about 1 to 12 1/2. A proper ration would be about 1 to 6 or 7, or less. To balance this up alfalfa can be fed better than anything else in California, for that is very rich in protein and the cheapest supply of protein that there is. If you give the cows a good supply of alfalfa hay with the green corn, you will have an ideal combination.

Dry Sorghum Fodder.

Is Egyptian corn fodder good for cows? I have been told it would dry up the milk. I have several acres and would like to feed it if it is not harmful.

Dry sorghum fodder is counted about the poorest roughage that one would think of harvesting. It is much less valuable than Indian corn fodder. Egyptian corn is one of the non-saccharine sorghums which are valuable both for grain or for green feeding. We never heard of direct milk-drying effect, though such a result might be expected from feeding such innutritive material, which is also difficult of digestion. If fed for roughness it should be in connection with concentrated foods like bran or oil meal or with green alfalfa. No cow can give much milk when the feed is hardly nutritive enough to keep her alive.

There seems to be, however, much difference in the dry fodders from different varieties of sorghum. One grower writes: "Kaffir corn is the only variety within our knowledge of which the fodder is of much value. We consider the fodder much more preferable than that of the ordinary Indian corn, and our stock eat it much more readily than the sweet sorghum. However, it requires a much longer season in which to ripen than does any of the other varieties, for which reason it is less desirable to plant in midsummer."

Steers on Alfalfa.

How much alfalfa hay will a two or three-year-old steer eat per day, and about what is the gain in weight per day?

A steer will clean up about 33 pounds per day. Steers will make about 1 1/2 pounds gain in weight per day.

Concentrates with Alfalfa.

I have a good supply of alfalfa hay and have been feeding this as a straight feed for my dairy cows. They are not, however, doing as well as they should and I am looking for some good feed to go with it.

You could probably get better returns by feeding about a pound of cocoanut meal and three of dried beet pulp than by any other combination of concentrates with straight alfalfa. If you are producing market milk or butter prices justify it, more concentrates could profitably be fed. It is an expensive proposition to build up a properly balanced ration with alfalfa and concentrates alone, and unless market milk is being sold, it usually does not pay. The cheapest way to provide a balanced ration is not by concentrates, but by wheat or other grain straw, and let the cows eat all they care for. This is very cheap and helps to balance a ration with green or dry alfalfa hay, is usually cheap, and is fine for cows. Both are much less expensive than concentrates.

Chopping Hay for Horses.

What saving may be made by chopping all oat hay when fed to horses?

There is no particular saving in chopping hay unless the horses are worked very hard and for very long hours, as is often the case with express horses in the cities, or unless the power for cutting is very cheap and feed high. The idea is that, except in unusual cases as above mentioned, the horses can do their own grinding cheaper than it can be done by power. Somewhat less hay is wasted when fed cut than when fed long, but if they are not fed too much long hay they will waste very little.

Grain for Horses.

What is the best formula for feeding work horses with oat hay, alfalfa, barley (crushed) and corn as rations?

Feed one-half oat hay and one-half alfalfa hay, about 1 to 1 1/2 pounds per day for each 100 pounds live weight of the horse. Add to this from 3/4 to 1 pound of rolled barley or corn for each 100 pounds live weight. If the corn is on the cob, four-fifths of its weight is corn; that is to say, 5 pounds of corn on the cob has 4 pounds of grain.

Feeding Cut Alfalfa Hay.

Would alfalfa hay, cut, say, from one-half to three inches in length be better than whole hay for hogs, cattle and horses, and if it is better, should it be fed wet or dry?

Cattle and horses do much better when fed chopped alfalfa hay than when fed whole hay. They can eat the required amount in much less time and with less exertion. For cattle and horses the hay should be cut about one inch long and fed dry. There is no advantage in chopping alfalfa hay for hogs unless it is mixed with ground grain and made into slop. - L. P. Denny.

Storing Cut Alfalfa Hay.

We are planning on cutting our next season's crop of alfalfa with a feed cutter and storing it in a barn for winter feeding.

The hay must, of course, be thoroughly cured, because of the great danger of heating in a tight mass. A. Balfour says: "I have been cutting alfalfa into a barn for wo seasons. It is absolutely necessary to have the sides and floor tight, and it is easier to feed it if it is in a loft. The hay is best stacked first, and must be thoroughly cured."

Alfalfa Grinding.

Is the curing of alfalfa for grinding different from ordinary; has it to be chopped before grinding, and what is the cost of grinding?

Alfalfa hay should be cut when the very first blossoms commence to appear. At this point the plant contains the greatest amount of protein; from that time on until seed time, the protein diminishes and fiber increases. To make meal, hay should be well cured, have gone through the sweat, and should be dry, or as near dry as possible. It mills easier when dry and makes a finer product. It should be cured so as to retain the green color. To grind it, it is not necessary to cut it before grinding, it mills better if ground just as it comes from the stack. The cost of milling hay varies with the size of the machine, condition of hay, whether dry or damp, or whether tough or tender. With larger plants of a capacity of four to five tons per hour, it costs about 45 cents a ton to put it in the sack, exclusive of the cost of sacks; and with smaller, it runs from that on up to $1 to $2 per ton.

Feeding Calves.

How soon can calves be weaned and not hinder their growth? After weaning, what would you advise to feed them?

After the calf has once nursed, it should be taken away from its mother, but fed its mother's milk for a few days, depending on the vigor of the calf. Commence to add skim-milk after a week or ten days, adding a small amount at first and increasing it daily until the calf is on an entirely skim-milk diet. The milk must be sweet, it must be as warm as its mother's milk and the calf must not have too much of it. Four quarts at a feed twice a day is sufficient for the average sized calf for the first month, then increase it accordingly. Add a spoonful of ground flaxseed to each feed and teach the calf to eat a little grain as soon as possible. Ground barley is the most economical feed to balance a ration containing so much skim-milk. If calves show a tendency to looseness of the bowels, feed less milk, and when this does not remedy the trouble, heat some skim-milk to boiling and when it is cooled to a proper temperature feed this to the calf. A good grain ration to feed calves along with skim-milk is ground barley with green alfalfa hay. When the milk is cut off, feed barley and bran soaked with molasses water. Put a pint of molasses in a pail of water and dampen feed with it. This amount will dampen three bushels of feed. - W. M. Carruthers.

Winter Feed for Sheep.

What would be the best to sow for sheep pasture - barley, oats, rye, vetch or rape?

Of the grains, rye is usually found to be best for quick winter growth, and rye and vetches sown together are very satisfactory, because the rye holds the vetches up so that the whole growth can be more successfully handled with the mower, and if grown that way and fed green in a corral, a very large amount of good feed can be secured. Sufficient experiments have not yet been made with rape to fully demonstrate its value. Even if it grew well, it would be inferior in nutritive value to vetches and rye.

Balanced Rations.

What is a balanced ration for milk cows and brood sows?

When plenty of alfalfa is available many dairymen feed that alone. It is better to feed a little corn, grain hay, beet pulp or the beets themselves to balance up the ration. Some of the best concentrates to feed to offset alfalfa hay are ground barley and dried beet pulp. The same thing can be said about the sows. They will consume about 10 pounds of chopped alfalfa per day and all the skim-milk that is likely to be given them. Not more than eight pounds of concentrates need be fed, of which one-fifth may be bran, the same amount, or more, of cocoanut oil cake, and the rest corn or barley. With plenty of skim-milk and alfalfa, but little grain or other concentrates will be needed. A few beets will also go well with alfalfa.

Pasture and Cover Crop.

I am thinking of sowing burr clover with rye to be plowed under in the spring. Is it good policy to sow rye with clover?

Burr clover and rye would be very satisfactory for sowing, after the rains, to secure a winter growth for plowing under in March or April, or earlier if the growth should be large enough to warrant. Such a cover crop can be pastured lightly to advantage.

Cutting Corn for Silage.

What is the best time to cut corn for the silo? What length is it cut? Is water put on it when it is put in the silo?

The best time to cut corn for the silo is just as the kernels are beginning to glaze. It is cut with a proper ensilage cutter into half or three-quarter inch lengths. No water is used, unless the corn should be unusually dry, with shriveled leaves; in that case, the use of water to compensate for the loss of moisture in the stalks and leaves is desirable.

Fall and Winter Pasturage.

What do you advise for planting in the fall for winter pasture in the Sacramento valley? Are field peas suitable?

The common California field pea, called Niles pea, the Canadian pea, the common vetch (which is sometimes called the Oregon vetch because the seed is largely grown in that State) are all suitable for fall planting and winter growth because they are not injured by ordinary valley frosts. Aside from legumes, you can get winter feed from fall-sown rye, Essex rape or kale.

Summer Pasture for Hogs.

I want to pasture hogs in the San Joaquin valley this spring and summer. Have water for irrigation, but will not have time to get alfalfa started sufficient to pasture.

Sorghum can be planted with pumpkins or some root crop between the rows. The root crop or the pumpkins could be used in the later summer, while the sorghums could come between the natural grasses of the early spring and the root crops. A strictly pasturage scheme is to sow wheat or barley and turn the hogs on this, so that they will eat within certain prescribed limits. In order to do this, the field needs a shifting fence, so that the hogs can be driven from one section to another - never letting the hogs eat too closely, as they will kill off the stand.

Size of a Silo.

I am planning to build a silo 8 feet high and 10 feet across. Will ensilage (corn, oats) keep well in a silo of those dimensions?

The silo you are intending to build is too shallow, and would hold only a very small amount of silage. There would be several inches loss of silage before you could start feeding, and you would have to feed at least two and probably three inches off per day in order to keep the food from spoiling. Sixty inches of silage would thus only last about twenty days. Also, the deeper a silo is, the tighter the ensilage is packed and the more will be contained in a cubic foot. The following table will give suggestions as to dimensions:

Diameter. Height. Capacity. Diameter. Height. Capacity. 10 feet 25 feet 36 tons 14 feet 34 feet 115 tons 10 " 28 " 42 " 15 " 34 " 131 " 11 " 29 " 60 " 16 " 35 " 158 " 12 " 32 " 73 " 20 " 35 " 258 " 13 " 33 " 83 "

A cow can consume four tons of silage in 180 days and more or less as you care to feed, so by figuring out how long you will probably feed, you can see the size of silo to build at once.

Soiling Crops in California.

What are the dates for planting crops to be used for soiling in your State?

We are using Indian corn and sorghums of various kinds for soiling to a certain extent. There is also some cutting and carrying of alfalfa, although most of the alfalfa is pastured. Dates of planting depend upon the frost-free period; sometimes beginning in April, and successive planting for later growth as water may be available for irrigation. There are places where one can see standing corn and sorghum untouched by frost as late as December 1. In other locations the growth of these plants have to be made between May and September. We have also winter-soiling practiced to a small extent in this State and for that purpose rye and barley sown at the beginning of the rainy season are used to some extent.

Brewer's Grains for Cows.

Are sprouted barley grains that may be had from breweries good for milch cows? Will it increase the milk, or will it dry up the cows?

Professor Henry, in his standard work on "Feeds and Feedings," says: "Fresh brewer's grains constitute one of the best feeds for the dairy cow. She is fond of them and they influence most favorably the flow of milk. Fed while fresh in reasonable quantities, supplemented by bright hay or corn fodder for dry feed, the grains being kept in tight feed-boxes which can be kept clean, and with other conditions favorable to the healthfulness of the cow, no valid objection can be raised against this form of feed. From 20 to 30 pounds of wet grains should constitute a day's allowance."

Feeding Pumpkins.

What is the proper way to feed pumpkins to cows? Some say to cut them in halves; while others say they must be chopped fine enough so that the cows cannot choke on them. Some tell me the seeds tend to dry the cows up, and should not be fed with pumpkins.

Pumpkins should be either cut in halves or broken in large fragments so that the stock can get a bite at them or else should be chopped fine, and we could never see the advantage of going to that trouble. Cutting into medium-sized pieces is dangerous because of the temptation to swallow them whole and thus getting choked. It is not necessary to remove the seeds.

Feeding a Family Cow.

What shall I feed family Jersey cow in addition to alfalfa hay to insure a good supply of milk?

One of the best things to feed in addition to alfalfa hay is a couple of quarts of middling or bran twice a day, with which is mixed a cup of molasses with enough water to make a nice paste. Dried beet pulp is exceptionally good with alfalfa, if it is available, this also to be moistened before feeding.

Rolled Barley for Cows.

Will rolled barley hurt milk cows, say two light feeds a day? Will it not do about as much good as the same amount of bran?

Certainly not and otherwise will be good if not used in excess to encourage fattening. Bran is a better feed for milk because it has a higher protein content.

Horse Beans and Pie-melons.

Would it pay me to raise horse beans for fattening hogs? Horse beans do well. Would citrons do well there without irrigation, and would they be better than stock-beets for hog feed?

We do not promise anyone that anything will pay. Horsebeans are good with other feeds for hogs. Theoretically, they will balance well with pie-melons and beets, and both the latter will produce well on good land with proper cultivation in the valley you mention. Theoretically, also, we would rather have beets than pie-melons. The hogs will tell you the rest.

Horse Beans.

Are "horse beans" a leguminous crop and how does their feeding value for hogs compare to cowpeas and Canadian field peas?

They surely are legumes, and they resemble so closely in composition the other legumes which you mention that their feeding value would be practically the same.

Storing Stock Beets.

What is the best method of storing stock beets and stock carrots in this climate? We can let them remain in the ground and grow until February or March and would like to preserve them for feeding as long as possible.

Stock beets and carrots can be stored in California without recourse to covering with ground or use of a cellar. They keep very well during the winter if piled under cover in such a way as to keep cool and dry.

Kale for Cow Feed.

What is kale worth for cow feed as compared with alfalfa, also can it be cut and cured the same as alfalfa and what variety is the best?

Kale is very similar to cabbage in growth, and for feeding purposes. For cow feed it would have about three-fourths the amount of digestible nutrients as green alfalfa, but would have an added value on account of its succulency. It would go especially well with alfalfa hay. The Jersey or Thousand-Headed kale is considered the standard for stock or poultry feed. It is always fed fresh and is not made into hay.

What Kind of Beet for Stock?

Which would be most valuable to plant on river-bottom land for cattle and hog feed, sugar beets or mangels?

Grow a large stock of beet by all means - either a mangel or a tankard. Usually you will get more weight than with sugar beets; the cost of harvesting is far less, and the nutritive contents high enough.

Keeping Pumpkins.

What is the best way of storing pumpkins, under ordinary farm conditions, in a climate such as we have here in northern California? I have no facilities for cold storage.

All you have to do in this climate to keep pumpkins is to keep them out of reach of the stock. They do not need storage of any kind, but will keep in good condition during the late autumn and winter months in any open-air place where they may be convenient for feeding purposes. In parts of California where there is hard ground freezing, protection must be given by covering with boards or straw or any other material available. We have no need for root cellars or cold storage, for our winter temperatures are neither high nor low enough to hurt them.

Grape Pomace as Hog Feed.

What is the value of grape pomace as a hog feed?

It has been sold for 50 cents a ton as it comes from the press at the winery and when a person has not got any surplus of other feeds, it is evidently worth that and then some. The only way to feed it is to put it up in a big pile and let the hogs take it as they want it. It will help keep them growing through the winter provided they have other feed with it that might not be sufficient without the pomace.

Proper Feeding of Young Pigs.

If I put two 50-pound shoats to an acre of barley that will yield 10 or 12 sacks of grain, how many months could they be kept there to advantage, and what gain could I expect them to make in that time?

If the pigs have been properly fed and were of good stock, they should have attained a weight of 50 pounds at three or four months of age. Pigs in this condition would be more likely to lose than gain turned on a dry barley field, even if the yield were double what you state. Barley is an excellent fattener for mature hogs, but is a poor food for young growing pigs. Young pigs should have a balanced ration, which may be defined as a little of almost all kinds of feed and not all of any one kind. We have pigs running on a barley field such as you describe, and in addition to the barley we feed them once a day a slop composed of wheat middling and bran in equal parts by measurement, to which we add about 8 per cent tankage, and they seem to be moving along nicely. Without the slop we don't think they would hold their own. - Chas. Goodman.

Pie-melons and Pigs.

I have 14 sows which were fed almost entirely on pie-melons and milk, not much of the latter. Out of the 14, only 3 sows have saved any pigs; the rest lost all the young they had. Four or five sows that for the last three weeks have had no melons, nothing but green grass and a little whole barley each day, are saving their pigs all right.

Pie-melons are poor feed and pigs which are not given anything better ought to fail. "Green grass and a little whole barley" is much better feed than pie-melons. Pie-melons are useful fed with alfalfa hay or some richer food.

Wheat or Barley for Hogs.

Which would be the better grain for me to buy for hog feed; wheat at $1.30 per hundred, or barley at $1? Would it be worth paying 10 cents a hundred for rolling, and then haul the grain 8 miles by wagon?

Wheat is only considered about 10 per cent more valuable as a hog feed than barley, so that in your case, barley at $1 is the cheaper. In Bulletin 80 of the Oregon Station it was found that crushed wheat was 29 per cent more efficient than the whole grain, and it is safe to say that barley will run about the same, enough so at any rate to pay the extra 10 cents a hundred for crushing and the hauling.

Grain and Pasture for Pigs.

What is the most profitable amount of grain to feed to spring pigs while on alfalfa pasture, from the time of weaning to the time of marketing?

We doubt the profit of feeding whole grain to hogs of any age while on green pasture. On almost all kinds of land they will get enough grit to keep their teeth sore, hence they will not masticate the grain thoroughly. Perfect mastication is very essential. We would feed the pigs all the slop that they would clean up good twice a day. The slop to be composed of equal parts of corn, barley meal ground fine, and wheat middlings mixed with milk. There is nothing in all the world like milk for growing pigs. If milk is not to be had, we would add from 5 to 10 per cent meat meal, which we consider next to milk. If whole grain is to be used, it should be thoroughly cooked on account of the pigs' teeth not being in condition to chew the hard grain. - Chas. Goodman.

Growing Pigs on Roots and Barley.

We can raise all kinds of root crops, such as carrots, sugar beets, rutabagas, etc., and cow peas and pumpkins do wonderfully well. Will hogs do well an that kind of diet, especially if given a little barley with it?

The plants that you mention are good for hog feeding and can be used to advantage with a little barley as you suggest. None of these plants are, however, rich in protein as alfalfa and the other clovers are. The reason why we get such a rapid and satisfactory growth of young hogs in California is due to the fact that they are largely kept on alfalfa and rapid growth is the product of a sufficient protein content in the fodder. Both common field peas and cowpeas do not possess this element, and if you can grow them they will serve as a substitute for the other legumes, such as alfalfa. If you are feeding skim-milk, which is rich in protein, roots and grain will go well with that.

Wheat and Barley for Feeding.

What is the difference in the feeding value of wheat and barley for hogs and horses?

There is very little difference in the chemical composition of wheat and barley. In their physical condition there is much difference, chiefly because of the adhering chaff of the barley, which makes it more digestible because it separates the starchy mass and enables the gastric juice to work upon the particles more readily and quickly. Oats also have this character. This is very important in the case of horses, which can quickly be put out of condition by feeding wheat. For hogs and chickens it makes much less difference, and the absence of the chaff gives a greater amount of nutritive matter to the ton, so that wheat is worth more at the same ton price. But look out about giving horses too much wheat.

Part VII. Diseases of Animals

This division is largely compiled from the writings of Dr. E. J. Creely of the San Francisco Veterinary College.

Abscess of Parotid Gland.

My horse has had a bad cold and it has a large lump on its neck which keeps running and does not seem to get any better; it has been running for two weeks.

This horse has an abscess of the parotid gland and the abscess should be opened large enough so that the finger can be introduced to break down adhesions, so that proper drainage can be established, after which wash out with a 5 per cent solution of permanganate of potash. As this is a dangerous location for a layman to interfere with, owing to the branching of the carotid artery, pneumogastric nerve and jugular vein, it should be done by a qualified veterinarian.

Forage Poisoning.

Last fall one of our horses was taken ill and had a swollen jaw. He died soon and we supposed that he had been kicked and died of lockjaw. This spring another was taken ill. He began dragging around, making an effort to eat and drink, but not being able to swallow much. Something seemed wrong with his throat and his hind legs. In two or three days he got down, seeming to have no strength in his back. He kept struggling for two days, not being able to swallow much; so we put him out of his misery. Since then two others have gone off the same way.

The trouble is due to forage poisoning, caused by the eating food infested with poisonous moulds. The symptoms are inability to swallow (paralysis of the muscles of deglutition) and paresis of the hind and forequarters. When the symptoms become advanced, treatment is of little avail. However, further troubles can be prevented by ascertaining the food which is infested with this mould. Ofttimes, however, such food may be apparently clean to the eye. Make a complete change of food and a thorough cleaning of your stable and corrals of all old fodder which might be in the mangers, or in any accessible place. Very frequently old food which is left in the bottom of mangers becomes mouldy, and horses picking for grain which might be left in it, eat considerable quantities of this spoiled fodder, get poisoned.

For a Scabby Swelling.

One of my cows has a swelling on her hind leg with little scabs on it, first it was on the front leg. It is as big as your hand.

Use the following, applied once daily: Olive oil, 1 pint; turpentine, 2 ounces; oil cedar, 2 ounces; lysol, 1 ounce; mix and apply.

An Easement in Bloat.

What can be done for bloating?

It does not seem to be generally known that to put a bridle on a cow or put a stick in her mouth and tie tightly with a string or strap up over her head, so as to keep her jaws working, will relieve bloat. We have given common soda and salt with good results to our milk cows. Take a whip and run her around the corral, after giving the soda. This treatment causes the wind to pass off.

Fatal Skin Disease.

About two months ago a horse was turned out in pasture. Several of the horses in the pasture started to lose their hair. It seemed to fall away from the hide, and leave the skin exposed. The horse that was newly turned to pasture got the same disease and died. The other horses did not die. The hair on the horse that had died had fallen off from the sides and hind legs.

This is gangrenous dermatis, a gangrenout inflammation of the skin. It is due to mould, must or vegetable fungi. Remove to a new pasture, give food free from the fungi, and apply the following ointment to the skin: Lanoline, 8 ounces; zinc oxide, 1 ounce; Pearson's Creoline, 1/2 ounce; tannin, 3 drachms; mix and apply once daily.

Shoulder Injury on Mare.

A young mare that bruised her shoulder on the point with collar. It was lanced and now has a hard lump or callous, about three inches in diameter. What is best to do? She is not lame, but it would interfere with the collar.

Get a qualified veterinarian to operate and entirely remove the growth or you may use the following mixture to see if it will not cause it to partly absorb and then use a dutch collar or a specially padded collar: Compound tinct. iodine, 4 ounces; sulphuric ether, 2 ounces; oil cedar, 2 ounces; turpentine, 4 ounces; mix and apply once daily until blistered.

Horse with Worms.

What is the best remedy for a horse that has worms? I would like to know, as I have a horse that is getting poor with this trouble.

Mix 1/2 pound pulverized and dried iron sulphate and 1/2 pound bicarbonate of soda, and give one teaspoonful each morning until the medicine is gone. After the last dose give the following: Turpentine, 2 ounces; fluid extract male fern, 1/2 ounce; Pearson's Creolins, 1 ounce; raw linseed oil, 1 pint. Mix and give all at one dose. To improve the general condition one may give artificial Carlsbad salts, 1 tablespoonful in each feed, and each dose to have added to it 3 to 5 grains arsenious acid. If plenty rock salt is allowed for horses to lick, they will be protected against intestinal parasites to a slight but useful degree.

Is It Mange?

We have a horse five years old that is always scratching and biting himself as if he had mange or lice. He seems to itch more on his shoulders and front legs than any other place. We have washed him with a carbolic wash, also with a tea made from tobacco, but so far have been unable to stop it. He often bites his legs below the knees until he takes off all the hair and part of the skin. None of the other horses are, troubled, although this horse has been troubled for three years.

Apply the following: Lysol, 1 ounce; kerosene, 4 ounces; formalin, 2 drachms; cotton seed oil, 9 ounces. Mix and apply once daily after washing with hot sheep dip solution 10 to 100.

Horse with Itch.

For about a year my horse has been itching so badly that he has rubbed off all the hair on certain parts of his body. Lately he bites his tail.

Whitewash the stall once weekly, scrub the harness, brushes, combs and every stable appliance that he has come in contact with. Don't use the same appliance on other animals that you use on this horse. Use the following mixture once daily on affected spots: Milk of sulphur, 4 ounces; tincture of iodine, 4 ounces; turpentine, 4 ounces; kerosene, 16 ounces; cottonseed oil, 120 ounces.

For a Bowel Trouble.

What can I do to relieve a horse that balls up on alfalfa at the time of the first symptoms? I have been bothered considerably with this, and although I know the symptoms, I can never seem to relieve the pain before the veterinary is called.

Give the following prescription: Fluid extract Cannabis Indica, 3 ounces; sulphuric ether, 2 ounces; spirits turpentine, 3 ounces; oil peppermint, 10 drops; raw linseed oil, 24 ounces. Mix. Give one-half at once, balance in one hour. If not relieved give several hotwater soap-sud injections.

Abnormal Thirst of Horse.

I have a horse with an abnormal desire for water. I notice that in drinking she always wants more than the others. I also notice she perspires more freely in the harness and even will sweat in the barn at night.

Your horse has kidney affection, probably due to feeding hay rich in alkalines. Treatment: Change the feed and give 1 quart of thick flaxseed tea three times daily.

Scours.

Kindly recommend a treatment for a horse troubled with scours. He is on dry feed, but the trouble continues.

Give very little water mornings and while worked, but give plenty at night. Feed dry rolled oats, oat hay, one handful of whole flaxseed at night, and the following powder: Bismuth subgalate, 4 ounces; iron sulphate, dessicated, 8 ounces; bismuth subnitrate, 8 ounces. Mix, and give a heaping teaspoonful each morning.

Depraved Appetite.

I have a colt about one year old that continually delights in chewing up harness, ropes, chews on the manger and, in fact, anything it can get a hold of.

This is a condition caused by something being lacking in the system (lime, salts, etc.). Give plenty of salt, good food, grain, etc. Get this prescription: Iron sulphate, 2 ounces; soda syposulphate, 4 ounces; Gentian root pulv., 2 ounces; ginger, 1 ounce. Mix and give teaspoonful daily.

Good Dentist Needed.

I have an old horse which has always been fat and quite full of life until right lately. Now he is getting thin and looks bad. He eats his food all right. I had his teeth fixed a few weeks ago. The man said they were bad and he fixed them as well as he could.

There is probably an excessively long molar projecting into a cavity and the projecting molar should be cut off by a qualified veterinarian. The horse will begin to pick up and grow fat almost as soon as the condition is relieved. Most horse owners will permit every person with a float to ruin a horse's mouth without inquiring whether the dentist possesses proper qualifications as certified by a State license and diploma.

Kidney Trouble.

My horse has some trouble in passing water. What can I give him that may be put in the mash? I don't think his trouble is due all to old age, for it didn't come on gradually.

Give gran. sal nitre: a teaspoonful daily in water is good to stimulate the kidneys.

For Chronic Indigestion.

I have given my horse condition powders for indigestion, but her hair is rough still. Do you advise feeding on the road when a horse leaves the stable at 10 a. m., traveling continually for thirty miles, returning 5:30 p. m., being fed at 7 a. m.?

A great majority of condition powders contain resin and antimony. While a slight amount may be beneficial, continued use results in affection of the kidneys by over-stimulation. Give the following for indigestion: Bismuth subintrate, 1 ounce; powdered pepsine, 1 ounce; soda bi carbonate, 12 ounces; carbonate iron, 2 ounces. Mix and give a heaping teaspoon twice daily. By all means feed your horse three times daily and water as often as you can. It is unnecessary to warn you that the horse must not be overheated when you give the noonday feed.

Wound Sore.

My colt got its hind leg cut on barbed wire some weeks ago. There is a hole about an inch and one-half deep in the center of the sore which will not heal. The inside of the sore does not seem very tender, but the leg stays swollen all of the time and is somewhat feverish.

This is probably a fistulous track that should be curetted by a veterinarian, after which the following formula could be used to heal: Acetanilide, 1/2 ounce; zinc oxide, 1/4 ounce; bismuth subgalate, 1 1/4 ounce. Mix and apply on cotton and bandage once daily after washing.

Warts on Horse.

How can warts be removed from a horse's hide?

We use sulphuric acid. The results were favorable from the very start. The warts rapidly shrunk away and finally disappeared entirely. The acid is applied to the crown of the wart with a small swab or similar instrument, and only in sufficient quantities to wet the crown surface of the wart. It should be applied about three times a week until the wart is well reduced. Don't use too much acid, and don't keep up the application too long - A. F. Etter.

Kidney Trouble in Horse.

What is the remedy for a horse that stops often to urinate while working?

The horse is affected by an irritation of the kidneys. Give 1 quart of flaxseed tea daily, change the food and give 1 drachm of C. P. hydro-chloric acid in one bucket of drinking water.

Castration of Colt.

Which is the correct and best way to castrate a yearling colt, with an emasculator or a blade, and when is the proper time?

An emasculator is the only instrument to use in castrating. The object in using any instrument is to prevent a hemorrhage, and nothing works with so much certainty and quickness. The A. Hausman and Dunn emasculator is recommended. The proper time is when the weather is mild, the grass at its best and the colt in good condition.

For a Chronic Cough.

We have a mare seven years old that is troubled with a chronic cough, and at times shows symptoms of heaves, and also has occasionally a white foamy discharge from the nostrils. She is a greedy eater and drinker and her excreta is often very offensive.

If she expels flatus when she coughs, this would indicate a predisposition to heaves. Wet all food, as dry or dusty food aggravates the cough. Give the following: Spirits camphor, 4 ounces; Fl. Ext. belladonna, 2 ounces; neutral oil, 8 ounces; oil eucalyptus, 2 ounces. Mix and give tablespoonful three times daily.

Chronic Indigestion.

I have a mare eleven years old. Give her plenty of oats, hay, grain and a little alfalfa hay three nights per week and leave salt where she can get at it, but she is falling off and her hair does not lie down properly. She eats well and her system seems to be in good condition. Have had her teeth attended to so she chews her food well.

This condition is caused by the animal not being able to properly masticate the food. Have your dentist examine the mouth again, or you can carefully examine the feces and see if it shows whole grain, or long pieces of hay.

For Short-Wind or Heaves.

I have a mare that has something wrong with her wind. About six months ago I noticed her wind was not good and she had a slight cough, and about a week later, while working her, she seemed to choke down and almost died before she got her wind, and since then she sometimes takes those spells should she trot off briskly for a short distance.

Give two 3/2-ounce doses of Fowler's solution arsenic daily. Dusty or musty hay will aggravate the symptoms. Thoroughly shake out the dust and wet the hay. Feed hay only at night. Give the animal as little feed and water as possible before being put to work. Continue this treatment one month if necessary. The following is a case of experience with this treatment: For a remedial agent we began to use Fowler's Solution of Arsenic, in two teaspoonful doses at first. once a day, put in the water with which the hay was moistened. These doses were given for a few days, then skipped for a day, then continued for five or six days again. This treatment has been continued. At times when the trouble was most severe, giving a great spoonful at a dose, twice a day for two days, then stopping for a day or two, always being sure to mix it with the water which the hay is moistened, so that it shall be taken into the stomach very slowly. This course of treatment has served to so relieve the disease that nature has nearly or quite overcome it.

Side-Bone.

I have a 1500-pound 3-year-old colt with small brittle feet that has side bone coming on left front foot caused by driving him barefoot on the road two or three months ago.

A good blister of the following once every six weeks for three times will stop the side-bones from growing. Side-bones on a draft horse are not considered an unsoundness; in light fast drivers it is an incurable blemish causing lameness. Side-bones cannot be removed. Use this blister: Simple cerate, 4 ounces; cantharides, 3 drachms; bin iodide mercury, 2 drachms. Mix thoroughly and apply after clipping hair.

Fungus Poisoning.

One of my mares, every evening after a full day's work harrowing, stands for an hour or so with her head to the ground, shaking it frequently and not touching the feed till the spell was over. She does not seem to be any worse off, and in the morning seems to be in good shape.

This is due to a mold or fungus in the earth or hay. Let them have access to plenty of water during the day. In the morning feed give a handful of sodium hyposulphate.

Treatment for Horse's Feet.

The soles of the fore feet of a fine 4-year-old horse, weight 1350, are rather spongy and grow down faster than the hoof, sometimes causing slight lameness. He is not on soft pasture, but is stabled all the time. Now have bar shoes on him. What treatment do you recommend?

Use leather, tar and okum and a dish-shoe.

For a Cleft Hoof.

I have a horse with a cracked hoof. One hind foot has been in a bad condition, the other seems to be beginning to crack. Can anything be done by feeding or otherwise to toughen the hoofs and render them less liable to crack?

Apply the following: Honey, 2 ounces; yellow wax, 4 ounces; tar, 2 ounces; olive oil, 8 ounces. Melt, mix and apply once daily.

Stiff Joints.

I have a horse that was bruised on the ankle about two years ago. This is now producing an enlargement of the bone and stiffness of the joint.

Apply the following liniment: Sulphuric ether, 1 ounce; tinct. iodine, 1 ounce; pulv. camphor, 1 ounce; alcohol, ounces; turpentine, 2 ounces; oil of cedar, 2 ounces.

Treatment for Nail Puncture.

Our horse got a nail in his foot. It was a wire nail, rusty, entering about one inch from the point of the frog, and just puncturing far enough to reach a sensitive part of the hoof. It occurred six days ago; the nail was pulled at once, the hoof cut open, and thoroughly cleaned with turpentine (the first thing we could get), then later filled with iodine. Since then I have kept on a flaxseed poultice.

The treatment with turpentine and iodine was proper and should prove a success. If the foot becomes tender and inflamed, it will be because all dirt was not removed from the wound, and the poultice should be taken off, all foreign matter removed from the wound, and the treatment repeated. In case of similar accidents, other disinfectants could be used in place of turpentine or iodine.

Pregnancy of Mare.

Is there any way to tell when a mare is in foal? I have had a veterinarian and he could not tell me.

There is no very good way to tell whether a mare is in foal for some time. Practically speaking, the safest way to do is to have her bred every time she comes in heat until she takes the stallion no longer. Even then some mares will come in heat a couple of times after getting in foal. If the sexual excitement speedily subsides and the mare persistently refuses the stallion for a month, she is probably pregnant, though not surely so. Also if a vicious mare becomes gentle after service it is an excellent indication of pregnancy; likewise pregnant mares will very often put on fat rapidly after conception and will be unable and unwilling to do as hard work as before. Enlargement of the abdomen, especially in its lower third, with slight falling in beneath the loins and hollowness of the back are significant symptoms, though they may be entirely absent. Swelling and firmness of the udder, with the smoothing out of its wrinkles, is a suggestive sign, even though it appears only at intervals during gestation. A steady increase of weight (1 1/4 pounds daily) about the fourth or fifth month is a useful indication of pregnancy. The further along the mare is in gestation the more pronounced the symptoms become. In the early stages it is naturally much more difficult to detect, especially with the great differences in different mares. Cessation of heat and changes of disposition are about the best signs in early stages.

Diseased Uterus of Mare.

I have a brood mare that has given me two fine colts, but for the last two years I have not been able to get her with foal. She takes service and then refuses service for three or four months, and about the time I come to the conclusion that she is safe with foal she will pass off great quantities of mattery substance. I have had her thoroughly washed out with Lysol previous to breeding, but so far she has repeated this performance each time about three or four months after service.

This is a disease of the ovaries or uterus; perhaps mumification of a foetus. Irrigate with a normal salt solution (teaspoon salt to each pint of warm water) only daily. Insert the solution through the neck of the womb into the uterus. Give internally 1/2 ounce daily of Fowler's Solution of Arsenic.

Deep-Seated Abscess.

I have a mule which has a swelling on the throat about where the throatlatch touches. It just seems to be swollen hard and not sore. I am using caustic liniment to fester it so it will come to a head and I can open it, but the liniment does not seem to do much good. The mule is losing flesh and does not eat much.

This mule should be operated upon at once by a qualified veterinarian. The application of liniments or blisters are useless; the knife only will effect a cure. The fact that the mule is losing flesh makes the case serious.

Cure for Cocked Ankles.

I have a 4-year-old mare that has cocked ankles, and would like to know what treatment to give her.

Cocked ankles are due to an inflammation of the tendons back of the ankle and a drawing up or contraction in consequence. Put on heel calks one inch, no toe, to rest and relieve the back tendons from strain. Apply the following liniment at night, after which put on cold-water swabs and let them remain all night: Soap liniment, 8 ounces; tincture iodine, 2 ounces; oil cedar, 4 ounces; sulphuric ether, 2 ounces. Mix and apply once daily.

Dehorning.

Which is the best way to dehorn cows and calves?

The best time to dehorn cows is in the spring, before the fly season starts. It is best not to have a cow too far along in calf before dehorning, as she is very apt to lose her calf. It is also better to dehorn before your cows freshen, because when cows are milking and are dehorned they will go back in their milk a great deal for the first month after the dehorning has taken place. Calves can be dehorned by blistering the little buttons before they adhere to the skull. This is very simple and not painful. First clip the hair about the horns and wet the little loose button and apply caustic potash, in stick form, by rubbing it on the damp horn. Remember, this must be done before the horn adheres to the skull. Also remember not to use water enough to run the lye away from the button and rub until the skin reddens. Also, look out to keep your end of the potash stick dry or you may dehorn the tips of your fingers.

Paralysis During Pregnancy.

I have a cow that will freshen in a few days. About six days ago she seemed weak in her hind legs and on going downhill would drag or stumble for 10 or 12 feet, then catch herself and go on rather wobbly.

Pregnant animals about to bring forth their young sometimes show a paralysis or loss of power in their hind parts due to pressure of foetus. Nature corrects this after birth.

Bloody Milk.

What can be done to stop bloody milk?

Milk each teat in a separate glass jar, let stand to ascertain which teat the red specks are coming from, then milk the teats clean and inject the infected teat with equal parts of hydrogen dioxide and water. After a few hours inject 4 drachms of ferric chloride in 1 ounce of water. Then milk clean.

To Cleanse Cows.

My cows are healthy and calves all right, but seem to have trouble throwing the afterbirth.

Wash out twice daily with about 1 gallon of normal salt solution (teaspoonful of salt to each pint of warm water). Give internally the following powder: Pulv. gentian, 4 ounces; puv. slippery elm, 1 ounce; puv. charcoal, 1 ounce; pulv. hyposulphate of soda. 8 ounces. Mix and give a heaping teaspoonful twice daily.

Treatment for Caked Bag.

I have a cow whose udder is caked hard and has been swollen from the udder to the forelegs. This latter swelling has gone down by applying equal mixture of turpentine and lard, but the udder itself still remains hard. When first noticed, one teat caked, then another, until all four are caked alike.

Insert a milk tube and inject the following: Hydrogen dioxide, 8 ounces; tincture iron chloride, 1 ounce; water, 7 ounces. Inject into each affected teat. Apply the following externally: Camphorated oil, 8 ounces; tincture belladonna, 2 ounces; oil eucalyptus, 2 ounces. Mix and apply twice daily.

Garget.

I have a cow which gave rich milk all the time, but now every time I milk her some yellow, hard substance will come out instead of milk. First from one teat, then the next, and when I strain the milk the strainer will be full of hard yellow specks.

Your cow has undoubtedly been affected with garget. This milk should not be used. The condition is best treated by massaging the udder every day with camphorated oil. It will also be necessary for you to continue to milk her regularly until about six weeks before she is due to freshen, at which time you should proceed to dry her up.

Infectious Mastitis.

We have a 2-year-old heifer, which, two weeks before she was due to freshen, had a large udder slightly caked. Upon pressing the teat a discharge of blood issues from each teat.

This is infectious mastitis. It may be due to a bruise or blow or infection introduced through the milk duct. The first is most likely. Apply camphorated oil externally and inject into the affected udder some hydrogen dioxide (peroxide of hydrogen. - EDITOR.). After ten minutes, milk out again. Repeat once daily.

A Mangy Cow.

I have a milk cow with some trouble about her head, neck and shoulders, which causes her to rub herself enough to make raw spots and take off most all of the hair from the parts affected. The trouble has been standing for 18 months, but I have been using medicine at different times, which stops the rubbing, and the part will cover with hair nicely again, but in due time the trouble shows up again.

This cow seems to have mange or scabbies, which is caused by a parasite and is easily spread by contact to other cattle. It should be treated by two or three applications, ten days apart, of a hot solution of creolin, well scrubbed into the skin. The solution is made by mixing five tablespoonfuls of creolin in a gallon of hot water. The treatment should be applied pretty well over the body to cover all the affected parts, and needs to be repeated in ten days to destroy the younger generation. The sheds should be cleaned and whitewashed.

Irritation on Back of Udder.

I have a yearling heifer which has sore teats and blotches just back of her bag which seem to itch. Her mother had a sort of eczema on her neck. I fear her sore teats will spoil her for milking when she comes in next year.

The following treatment is advised: Drench with 1 pound of Epsom salts dissolved in a couple quarts of water. The sores may be treated by washing them with a 2 per cent solution of one of the coaltar disinfectants, such as creolin. After the sores have been allowed to dry naturally, a very little powdered calomel may be dusted thereon. Do this every other day for a few days.

Enlarged Gland on Neck.

I have a calf that has a lump on her neck, which appeared when she was two days old. The lump is getting larger.

This is probably an enlarged thyroid gland. Apply the following once daily for several weeks and let it alone unless it becomes too large or gets very soft, which is unlikely. Churchill's tincture iodine, 8 ounces; turpentine, 1 ounce; sulphuric ether, 2 ounces; oil aniseed, 1/2 ounce. Mix and apply once daily.

Lumpy Jaw.

Some of my cows have hard lumps on their jaws, or lumpy jaw. Can that be cured, and how?

This is Actinomycosis (lumpy jaw) and is due to ray fungi (actinomyces) which are found originally on plants which enter the body in various ways. The trouble usually appears in the upper or lower jaws of cattle, where it generally produces tumors of bone or soft tissues. For treatment give 1 1/2 drachms of iodide of potash in 1/2 pint of water daily for 14 days. Increase to 2 drachms for 14 more days, and then gradually decrease. Divide the tumor and insert gauze saturated with tincture of iodine for 4 days. In 8 days a visible improvement will be noticed,

A Neck-Swelling.

My cow has a swelling under her neck between her jaw bones about the size of a baseball and almost as hard. It is not attached to anything apparently, but largely suspended by the skin at the entrance to the throat.

Cut directly through the center of the enlargement, clean to the bottom, splitting it wide open. Clean it out with peroxide of hydrogen, after which saturate absorbent cotton with tincture iodine, pack in tight and sew the skin to hold it in place. Remove the dressing in 48 hours and wash with sheep dip (tablespoon to 1 quart of warm water) twice daily. This may be tubercular, or the result of foxtail, etc.

Cow Chewing Bones.

One of my cows is continually chewing bones. What can I do to prevent it?

Give the cow good clean hay; some root crop, cocoanut meal, bran or soy-bean meal. If the cow does not stop mix in the drinking water twice daily a little dilute hydrochloric acid. Also, have boxes arranged near feeding stalls which contain wood ashes, slaked lime and salt.

Swelling on the Dewlap.

I have a cow that has a large lump at the point of the breastbone, the dewlap. This lump is as large as a cocoanut, and was caused, I think, by friction against a low manger in eating.

Get equal parts of tincture of iodine and soap liniment and rub onto the swelling twice daily for a week.

Barren Heifers.

I have three heifers, 3 years old, which have run with the bull right along and have failed with calf; have had three different bulls to them; what can be done?

There is a possibility of contagious abortion causing these heifers to fail to breed. If this has occurred in the herd, the heifers are very apt to be affected. If apparently healthy, reduce me feed and make the heifers take considerable exercise to reduce flesh. Give each a dram of powdered nux vomIca and one-half dram of dried sulphate of iron once daily in a little feed. Breed to a healthy bull when the heifers come in heat.

A Sterile Cow.

I have a very fine Jersey cow. I have had her to the bull every month, and can't get her with calf.

In an isolated case of this kind there is probably some disease of the generative organs or some condition whereby the impregnation cannot occur even when the animal is bred. The ovaries may be cystic; there may be chronic inflammation of the womb and possibly the mouth of the womb was injured at last calf birth and the scar prevents its admitting the fertilizing cells. If possible, a veterinarian should make a careful examination of this cow in order to determine what the trouble is. However, this treatment may be tried: About the time of coming in heat, give the cow a large dose of glaubers salts (one pound) and the nux vomica and iron treatment advised for "Barren Heifers" in another paragraph. Before breeding the cow, apply a little extract of belladonna and glycerine to the mouth of the womb and breed a few hours after.

Supernumerary Teat.

On the upper part of one of the hind teats of a young Jersey cow that freshened recently for the first time, there is a small growth from which the milk comes more plentifully than from the natural opening below. How, if at all, can this opening be closed without drying the cow? The milk from it runs all over the milker's hand and makes milking very disagreeable.

The only thing that can be done until the cow is dry is to tie the small teat up before milking. This can be done with a string, rubber band, or an ordinary clamp. If it is so small that the opening cannot be tied, there is nothing to do, except, perhaps to use, her as a nurse for calves. Two of these might run with her at a time, making way for others as soon as they are able to look after themselves. Quite a number of calves can sometimes be handled in a single year by a cow affected this way and the benefit to the calves might be nearly as much as by using the cow for butter production. When the cow is dry the teat can be amputated and the opening will close when the sore heals, or a stick of lunar caustic can be inserted into it, causing a wound that will heal solid.

Infection of Udder.

Last year one of my cows had milk fever which affected her udder. This year after freshening she milked two months when she suddenly went dry on one side of her udder. She is now badly stiffened up in her hind quarters and off her feed.

The cow has infectious mastitis due to introduction of some infection. Give a saline purge (1 pound. glauber salt), inject peroxide of hydrogen, after which pump in, sterile air. Apply externally camphorated oil once daily. Camphorated oil has a tendency to dry up the secretion of the gland and is used advisedly.

Lumps in Teats.

My cow has hard lumps in, her teats and lower part of the bag. These cause pain to her on milking, but there are no other symptoms of disorder. This condition has prevailed several months.

Give 1 drachm. iodide potash daily for one week; 2 drachms the second week 3 drachms the third week, add reduce as you began. If tumors are small and interfere with the flow of milk they can be removed.

Wound in Teat.

I have a cow with an open slit about one-fourth to one-third of an inch in the side of one teat. I have lacerated the edges and stitched the slit well together many times but the milk will ooze out and prevent healing together. I have used numberless milk tubes to no avail, as the flange on the tubes loose out. When I remove the flange the tubes creep up into the udder and it is a trouble to get them out again.

Wounds of a quiescent udder usually heal, but if the cow is in milk and the lesions involve the teats it is exceedingly difficult to heal the wound, as the irritation delays or interrupts the healing process. The following lotion is one of the very best to use for teat wound: Tinct. iodine, 2 ounces; tinct. arnica, 2 ounces; glycerine, 2 ounces; comp. tinct. benzoine, 2 ounces. Mix and apply twice daily after washing with 5 per cent solution carbolic acid and castile soap. Your milk tube must be an ancient one as all milk tubes of today are self-retainers and could not slip into the udder. Care must be taken to boil the tube previous to each using as you may cause an infection of the udder by a filthy tube.

Injury to Udder.

I have a cow which has a gathering in the back of her udder which seems to be some sort of injury. It has been there but a few days.

This injury was caused by a blow or traumatism. Thoroughly scrape out the diseased tissue and after washing with sheep-dip water (tablespoon to one quart) apply the following powder: Mix the following powder and apply it to the wound: Iodoform, 1 drachm; boric acid, 1 ounce; alum, 1/2 ounce; zinc oxide, 1/2 ounce. Be sure and insert this powder into the bottom of the wound, so that it will reach all diseased parts.

Blind Teat.

What can I do for a "blind teat"? The cow has just freshened and that quarter of her udder is very full, but there is no milk in the teat. I have been rubbing and greasing the udder. The blind quarter is slightly inflamed.

An artificial opening should be made in the teat at once. Call in the nearest physician unless you have a regular graduate veterinarian near.

Cow Pox.

I have a yearling heifer which is in fine condition and making good growth. But all four of her teats have sores on them and are mostly covered with scabs.

It is probably cow pox. Give a physic of glauber and epsom salts mixed 4 ounces of each to the heifer and double the dose to the cow. Apply externally, once daily, after washing, the following prescription: Zinc ointment, 4 ounces; iodoform, 1/2 ounce; glycerine, 2 ounces; carbolic acid, 2 drachms. Mix thoroughly and apply. to sores.

Cause of "Loss of Cud."

About three months ago a pure-bred Jersey commenced to fail on her milk and soon went dry, although on good feed. Did not seem to be sick, but did not eat ravenously as she generally did, and little was thought of it. During the past six weeks she has failed rapidly. Does not chew her cud, froths at the mouth, runs at the eyes, and when she eats anything much it bloats her. In fact, she seems bloated all the time. She is lifeless and will hardly move around, getting very thin, and hair standing the wrong way. Is there such a thing as a cow losing her cud?

Most people imagine a cow's cud is something material. As a matter of fact, in a certain sense the words appetite and cud are synonymous. You can say a cow has lost her appetite or a cow has lost her cud. Now, any sickness severe enough will cause a cow to lose her appetite. The bloating is caused from indigestion secondary to some organic disease, probably tuberculosis. Keep up the cow's strength by giving condensed floods or drenches of egg-nogg, gruel or greens. Give warm salt-water injections twice daily and give the following mixture: Quinine sulphate, 2 ounces; Antipyrine, 1 ounce; ammonia muriate, 3 ounces; alcohol, 1 quart; water 1 quart. Mix; give 2 ounces every four hours.

Calf Dysentery.

I would like to know the reason for bloody discharges from the bowels of a young six-day-old calf. There is a looseness of the bowels and the blood is intermingled with the excrement. There is not a profuse amount of blood, nor is it very dark in color, and it seems to be accompanied with mucus or light, thick substance.

This is dysentery, due to scours so prevalent in calves. Give 6 ounces olive oil, 4 drachms bismuth subnitrate and 1 drachm Pearson's creoline. The discharge is very dangerous to other animals.

Bovine Rheumatism.

Our Jersey cow got somewhat lame one year ago in one hip or leg after calving but soon got better. Last June when she came in one leg was lame. It seems to be in the stiffle joint and the first one above. When she walks she gets real lame.

Rheumatism is the trouble here. Give the following powder: Soda salicylate, 3 ounces; salol, 2 ounces; pulv. gentian root, 2 ounces. Mix and make 24 powders. Give four daily. Apply Pratt's, a good veterinary liniment.

Bleeding for Blackleg.

I have read several articles on blackleg, and it seems strange to me that no mention is made of an operation that is an absolute preventive, namely, bleeding in the feet.

The reason that no special mention of bleeding is made is that it is not now considered the preventive that it once was. Some people appear to have fair success with it, and others no success at all. The Bureau of Animal Industry states that the evidence indicates that bleeding, nerving, roweling or setoning have neither curative nor protective value and, therefore, should be discarded for vaccination which is now widely used as a preventive.

Poor Feeding, Depraved Appetite.

I have three cows. They have been fed alfalfa hay all winter and are in very good condition and seem otherwise in good health, and have salt to run to. Every time they chance to come to the yard they will pick up on old bone and chew it for perhaps a half hour. I always take the bone away from them when I discover it.

These cows have a depraved appetite, owing to the fact the tissues of the body are crying out for something lacking that is required in the system. Administer the following powder; also put a lump of lime in the watering trough: Pulv. gentian, 1 ounce; pulv. elm bark, 2 ounces; pulv. iron sulphate, 1 ounce; pulv. bicarb. soda, 4 ounces; pulv. aniseed, 2 ounces; pulv. red pepper 1/2 ounce; pulv. oilcake meal 10 pounds. Mix thoroughly and give a tablespoonful in scalded grain once daily.

Cows Swallowing Foreign Substances.

We recently lost a valuable cow, and when we opened her we found a large tumor or abscess at the top of the heart as large as a gallon jar. What caused it, or is there any danger of other cows taking it, and if so, what can we do?

This is a common disease among cows and is called traumatic pericarditis. The trouble arises from the habit of the cows picking up foreign substances such as wire, nails, or hairpins, and swallowing them. They are taken into the paunch and the digestive movements of this organ cause the foreign body to penetrate the lining and enter the heart, where it gradually causes death as it enters deeper. It is very common to find nails, etc., in the stomachs of old dairy cows which are killed at the slaughter-houses. If you had examined the animal carefully, you would find that some foreign body had penetrated the heart and caused death. There is no danger of any contagion arising from your cow.

Defective Urination.

I have a cow that seems to be in good health and gives plenty of milk. Nearly every morning when she is being milked she seems to want to urinate and will stand letting the water drip from her.

This trouble often results from the cows eating alkaline hay. Give her two quarts of flaxseed tea daily. Mix it with her food in which there has been placed one-half teaspoon of powdered Buchu.

Infectious Conjunctivitis (Sore Eyes).

I have several cows and heifers that are affected with sore eyes. The disease first makes its appearance by excessive watering of the eyes; then the center or pupil becomes white and later turns red of bloodshot.

Bathe thoroughly with the normal salt solution (teaspoon salt to 1 pint warm water), after which place in the eye and all around the mucuous membrane of the eye the following: Twenty-five per cent solution of argyrol, one-half ounce; apply thoroughly once daily and keep out of the sunlight if possible. Another treatment is: Bathe the eyes once daily with boracic acid 1 teaspoon, water 1 pint, after which thoroughly saturate the eyelids and eyes with 1 to 10,000 solution of bichloride of mercury. You are dealing with a disease that will spread throughout your herd if you do not take proper means to separate the affected from the well ones.

What to Do Against Tuberculous Milk.

I should like to know what could be done with a dairy where cows are dying with tuberculosis and the owner knows, but is selling the milk.

The case should be reported to F. W. Andreason, Secretary of the State Dairy Bureau, at San Francisco, for investigation by an inspector. If conditions are found as represented, the sale of milk will be prevented, as it is contrary to State law to sell milk from sick cows. County boards of health have also authority to prevent the sale of such milk in the county on the ground that this is a menace to the public health.

Effects of Ill-Feeding Pigs.

I have a couple of pigs, out of about 75 head farrowed last spring, which seem to have the staggers. They are looking fairly well, feed well on pasture and at feeding time are right there making as much noise as the others. They run around as if they had a shot too much.

Your pigs are suffering from acute indigestion, undoubtedly due to improper feeding. Cut down the rations, especially if they are getting grain. Give sick pigs two tablespoonfuls of castor oil each.

Sore Eyes in Pigs.

What is the matter with young pigs when their eyes swell shut? Before they shut they look as if there was a white milky scum over them.

There is some infection present, and a good cleaning up in needed. The sows and pigs should be dipped in a warm solution of some coal-tar disinfectant, and the quarters thoroughly cleaned and disinfected or changed to a dry warm place. The pigs' eyes should be washed with warm water and a few drops of the following solution dropped into eyes once a day for a few days: Have druggist prepare a 1 per cent solution of silver nitrate. After applying this the eyes had better be washed a few minutes later with water to which a little common salt has been added.

Hog Cholera.

I have a number of pigs which have been ailing for three weeks or so. They discharge a yellowish kind of manure at times, running of the bowels. The most striking symptom seems to be a partial paralysis of the hindquarters. The hogs will be walking along and seem to lose control of their hing legs. It seems to be spreading to the other hogs and a number have already died. Their appetite is poor.

This is undoubtedly hog cholera. The owner should appeal to the Experiment Station at Berkeley for serum and treat all well hogs and clean up as thoroughly as possible. The matter should also be reported to the State Veterinarian at Sacramento.

Pneumonia in Pigs.

What is the disease which may be said to confine itself, with few exceptions, to young pigs weighing 100 pounds or less? Its symptoms are at first sneezing and a mild cough. These quickly change to hard coughing and labored breathing, which as the disease progresses shows evidence of much pain. The appetite is lost and the eyes become gummed and inflamed. In some cases the pig lingers on for weeks, while in others death occurs almost immediately. Vomiting sometimes occurs.

It is pneumonia and in its treatment "an ounce of prevention is worth a pound of cure." Once pneumonia gets a foothold in a hog, the chances are so strongly in favor of death that recovery may be considered out of the question. Since remedies are not certain in the cure of pneumonia, it will be found that the prevention of the disease is the only real way to combat it. The main causes of the disease are exposure to draughts, sudden changes in temperature, damp beds, manure heaps as sleeping quarters, and exposure to the disease itself. Pigs in thin condition or weak constitutionally are more liable to contract the trouble than pigs in good flesh and healthy specimens. Good, dry, warm, comfortable sleeping houses, well ventilated and so arranged as to prevent crowding and piling up, will, I think, do more to prevent pneumonia than any other one thing. Some such preparation as advocated by the Government for the prevention of hog cholera will help keep the stock in a good healthy condition, the better to combat exposure. It is the little attentions that keep the herd healthy and in a vigorous condition, and by using simple preventatives, remedies will he found unnecessary. - H. B. Wintringham.

General Prescription for Hog Sickness.

My hogs seem to be mangy and scabby, but am unable to find any lice on them. They eat well, but vomit a good deal and are falling off in flesh.

They may be affected with a chronic type of cholera, and this should be determined by some one who can see the hogs. Make a general cleaning up of the hogs and quarters, using a dip and repeating in ten days. Hogs have a true mange as well as other animals. A change of feed may also be needed, depending on what is being fed and how the hogs are managed. Green alfalfa pasture with a moderate feed of shorts or middlings of wheat and ground barley made into a slop would be a good ration. Evidently there is some digestive trouble here, and a dose of croton oil (3 drops) mixed in a teaspoonful of raw linseed oil for each hog would be beneficial. Charcoal, ashes, salt and a little epsom salts would be of benefit to tone the digestion. The oil should be carefully mixed in the slop.

Pigs Out of Condition.

Of a litter of pigs weaned about a month several of them have itchy scabs on their legs, ears and noses, and those having white feet show reddish spots through the hoofs. They did not get it until after they were weaned. They are fed on soaked whole barley and have alfalfa pasture.

Put the pigs on a slop composed of wheat middlings and barley ground fine, with the hulls removed, and milk, or, in the absence of milk about 8 or 10 per cent of meat meal to which add some good stock food. Dip them with some standard brand of dip or apply crude oil to be sure that they were free from lice, fleas, etc. Give them good, clean, comfortable sleeping quarters and trust to nature to do the rest.

Paralysis of Sow.

During the last few days one of my sows appears to be paralyzed in her hind quarters and now cannot use her hind legs at all. She is about a year old and is due to farrow her first litter in and about six weeks.

It is paralysis due to advanced pregnancy. Give 4 ounces castor oil and 4 ounces olive oil. She will recover after parturition.

Rickets in Hogs.

A fine boar, 16 months old, weight about 380 pounds, well built, with little surplus fat, until lately has been very thrifty, but appears to be losing control over his legs. Can't step over the smallest stick without falling forward and acts like a foundered animal. He carries his back rather arching since this trouble came on. During my absence from home a hired man gave this boar a good beating with a pick handle, and it appears to have been the beginning of his troubles.

This disease is Osteo Rachitis (rickets). The abuse has probably aggravated the symptoms: This condition is due to a lack of hardening principles in the bones. Give 4 ounces of cod liver oil daily and plenty of lime water to drink. It will be all right to use him for breeding when he recovers. In addition to good food and pure water give daily a handful of a mixture of principally ashes and burned barley (charcoal) with the usual addition of salt, sulphur and soda. This mixture is good: Pulv. dried, iron sulphate, 4 ounces; soda bicarbonate, 8 ounces; soda salicylate, 2 drachms; pulv. aniseed, 4 ounces. Mix and give one-half teaspoonful twice daily.

Pigs Losing Tails.

We have five pigs, 17 days old, and when they were farrowed they had rings around the roots of their tails, and now their tails are dropping off.

This is caused by interference with circulation before birth. Apply tinct. iodine around the affected parts once daily and if it shows no signs of improvement after one week amputate.

Over-Fat Sow.

My brood sow is awfully fat; how should I feed her so that she don't get too fat? She is bred and it will be her third litter. She was running in the vineyard all winter, and I fed her a handful of barley every day or a few potatoes. Now she has free access to my growing barley field, and I give her half a dozen potatoes every day.

You need not worry about getting her thin. She simply requires less food. An animal excessively fat brings forth an inferior offspring.

Musty Corn for Pigs.

Would Egyptian corn that has been musty and then dried in the sun be fit for pigs? It heated and musted quite a good deal, but is dried well. The idea is, to grind it and then feed it in milk if good.

It is very dangerous to feed any stock moldy or musty food, especially pregnant animals. It is this kind of food which causes a majority of the abortions. Mold or smut in food is poisonous both to man and beast. It is usually almost impossible to get out of feed because it runs throughout the structure of the hay or grain.

Wounds and Wound Swellings.

What is the proper treatment for a fresh wire cut on a horse? How should saddle galls be treated? Is there any way to make the hair come in its natural color where saddle galls have been? How can an enlargement of a colt's leg, caused from a wire cut, be reduced?

After all foreign matter has been removed from a lacerated wound, like that made in a wire cut, the wound should be carefully fomented with warm water, to which has been added carbolic acid in the proportion of 1 part to 100 of water. It should then be bandaged to prevent infection. Zinc ointment would be a good thing to use under the bandage. For a simple saddle, or harness gall, some ointment like the following should be applied and the wound rested up: One pint alcohol in which are shaken the whites of 2 eggs; a solution of nitrate of silver, 10 grains to the ounce of water; sugar of lead or sulphate of zinc, 20 grains to an ounce of water; and so on. Or advertised gall cures may be applied. If a sitfest has developed, the dead hornlike slough must be cut out and the wound treated with antiseptics. There is no way we know of to make hair come in with natural color after a wound. The swelling on the colt's leg may he reduced by rubbing it well several times a day and at night rub in some 10 per cent iodine petrogen.

Fly Repellants.

Can you tell me what to use as a spray to kill the flies in my stable? In the early, morning the ceiling and sides are thickly covered with the pests partly dormant but not enough so that they can be swept down and killed. What spray can I use that will destroy them?

It is difficult to kill flies by spraying them. You can, however, spray the sides and ceiling of the barn with a spray of epsom salts (sulphate of magnesia) using about a cupful to the gallon, which will prevent them from gathering there. And since prevention is better than cure, flies can be kept from gathering around by, destroying their breeding places, if those are under one's control, by having all manure and litter removed before the flies have a chance to develop. The following may be found useful to readers as a spray to keep away flies: Fish oil, 2 quarts; kerosene, 1 quart; crude carbolic acid, 1 pint; oil of pennyroyal, 1 ounce; oil of tar; 10 ounces. Mix thoroughly and apply in a fine spray. The following has been successfully used to repel flies from cows: Nitro benzine, 5 ounces; carbolic acid, 3 ounces; kerosene oil, 3 ounces; sol. formaldehyde, 1 ounce; fish oil, 1 1/2 quarts. Mix and just touch the hair with the mixture.

To Destroy Fleas.

My barn, is full, of fleas I tried to destroy them by using cresodip, but did not kill them, all.

Fleas can only be permanently checked by destroying their breeding places which are in the dust! and dirt that accumulate in cracks and corners around barns, sheds and dwellings. Follow the cleaning up with a thorough distribution of flake naphthalene. This is most effective where the stable or room can be closed tight for half a day, or even 24 hours; An ingenious suggestion is made that if a sheep can be let run in and around the buildings where the fleas breed, they will soon be less numerous and as new batches hatch out the sheep will soon get them picked up, and after a while the place will be entirely free of them. But the sheep must be allowed to run all around the sheds and breeding places, as the flea jumps up, gets into the wool, and can never get out again. A hog can also be used as a flea trap. One reader says: Pour a little of the crude oil on the hogs' heads and along their backs, about a gill on each hog; This would run down the sides of the hogs and kill all the fleas on them. The oil also remains on the hogs for several days, and all the fleas that jump on the hogs from the ground stick fast and never jump off again. In about three weeks the fleas all disappear and the hogs look fine and sleek from the use of the oil.

Part VIII. Poultry Keeping

Largely compiled from the writings of Mrs. W. Russell James and Mrs.
Susan Swapgood.

Teaching Chicks to Perch.

What is a good method of breaking in young brooder chicks to use the roosts?

At from six to eight weeks old the chicks should be taken from the brooder quarters to the colony houses and range, or wherever they are to be located, and at this time they should be taught to perch. Have the new quarters arranged with low wide perches (1 by 3-inch scantlings); also make slatted frames by nailing lath or other such narrow strips two inches apart. Set these frames against the wall so that they will extend slant-wise under the perches, and have the corners on the other side of the room cut off by nailing boards across them. The chicks will run up on the frame to find a huddling corner and land on the perches, as they cannot rest on the open slanting frame. A little care for a few evenings in putting up those that remain on the floor and straightening them out on the perches will teach them the ropes. Where there are but a few to be taught, all that is necessary is to provide the low wide perches and shut out the corners, and a few of the smart ones will soon take to the perches, and gradually others will follow until all will be roosting.

Liver Disease.

I have hens which seem well in every respect up to the time of their combs changing color, when they die within three days. The combs turn a faint yellow, almost white; they are heavy, have their usual appetite up to the lost 24 hours. I have treated by giving small doses of castor oil and Douglas mixture in the drinking water, feeding on dry mash with plenty of green feed. There is no tendency to lameness nor limp neck. The droppings are loose and very white.

The fowls were victims of jaundice, which is a form of liver disease and caused by over-feeding on rich starchy foods that also cause fowls to become overfat. However, at the end of the laying season and the beginning of the molt the poultry keeper will lose some hens, even when kept under the best conditions, and especially hens of that age. In doctoring such cases in the way described, if the fowl does not improve in a couple of days, the hatchet cure is the most profitable.

Rupture of Oviduct.

I have had two other hens die suddenly when on the nest. The second one - we opened and found one egg broken near the vent and another with shell formed ready to be laid.

Rupture of the oviduct was probably the cause of the hens dying on the nest and is due to the same condition in the hens; that is, the straining to expel the egg necessary in the engorged condition of the internal organs from overfatness.

Melons for Fowls.

Have "stock melons" or "citrons" any merit as a green food for laying hens? Are the seeds of the above injurious to hens or cows?

Stock melons are desirable for chicken feeding if other succulent materials are scarce, but they are inferior to alfalfa and other clovers. Seeds are not injurious to stock unless possibly one should feed to excess by separating them from the other tissues. If melons are fed as they grow, no apprehension need be had from injury by seed.

Rape and Vetch for Chickens.

What time do you sow rape and vetch and are they good for chickens?

They surely are good for chickens or for any other stock that likes greens. They are winter growers in California valleys and should be sown in the fall as soon as the land is moist enough to keep them growing, or just as soon as you can get it moist either by rainfall or irrigation. Neither plant likes dry heat or dry soil.

Preserving Eggs.

What is a good way to preserve eggs for home use?

In a cool cellar, eggs will keep very well in a mixture of common salt and bran. Use equal parts, mix well, and as you gather the eggs from day to day pack with big end down in the mixture and see that the eggs are covered. Waterglass eggs are good enough for cooking purposes, but when boiled anyone that knows the taste of a strictly fresh egg can tell the difference in an instant; when fried the taste is not so pronounced, but it is there just the same; besides, when broken, they are a little watery. This watery condition passes off if left to stand for a few minutes. The best way is to use the waterglass method, is one quart of waterglass to ten quarts of water. Boil the water and put away to cool, when cold add the waterglass, mixing well, and store in 3 or 5-gallon crocks in a cool place. They will keep six months if good when put in. In all cases the eggs must be gathered very fresh, for one stale egg will spoil the whole lot, so great care is needed.

Dipping Fowls.

How do you dip hens to kill lice?

To dip fowls you must have a very warm day, or a warm room where you can turn them in to dry. I have know people to use tobacco stems, but it requires good judgment as to the right strength to use. The dips usually sold already prepared are safer, in my opinion, because they give directions as to quantity. Get a can of "zenoleum" or "creolium" - either is good - and have the water a little over blood-heat to commence; be very careful that the liquid does not get in the fowl's throat. If there are no directions with the cans, put enough in to make the water quite milky and strong smelling. It is best to make the hen sit down and with a sponge wet the back and head thoroughly, then under the wings and breast; if there are nits, don't be in a hurry to take the hen out, but let the dip get to the nits and skin on the abdomen. If the water is too warm it will be dangerous, as some fowls have weak hearts; that is the only danger, providing you dry them quickly.

Cure for Feather-Eating.

What is the cure for feather-eating?

Feather eating is the result of idleness or a shortage of green feed. The best way to cure it is to furnish the fowls with exercise. Boil some oats until soft, and when cooked stir in salt enough to taste and about a quart of good beef scrap; feed this for breakfast several mornings together. Make them scratch for the rest of their food in deep litter and give them sour milk to drink if you have it. If sour milk is not available, put a tablespoonful of flowers of sulphur in the boiled oats. The object is to cool the blood and furnish exercise. See that the fowls are supplied with mineral matter, such ash shells, bone meal and some, sand if it can be had. It is surprising the amount of sand that chickens will eat when carried to them in yards, so there must be a necessity for it, and if they cannot get to it, it pays to carry a good box full once in a while.

Cannibal Chicks.

What can I do to cure my chicks of eating each other?

Some kind of animal food is necessary when the chicks begin to pick toes, wings and vents. But the meat must always be cooked, the least bit of raw meat drives them wild as does the blood they can bring on each other. For that reason a strict watch must be kept to detect any case before blood is brought. Remove all weak chicks as they always go for the weakest, and as soon as one chick is picked on for a victim, remove it at once. Some people paint the toes with tar or liquid lice paint, but I have had the best success with bitter aloes mixed with water. A nickel's worth covers a lot of toes. It is best to buy a powder, then dissolve in a little water and paint wings, vent and toes. They won't take many pecks at them when they find they are so bitter.

Sunflower Seeds for Poultry.

What is the food value of sunflower seed as a ration for fowls, mostly laying hens? Should it be fed whole or crushed?

Sunflower seed is rich in oil, having the same proportion as flaxseed; otherwise it rates in value the same as grain. A little, not too much, fed whole is well relished by fowls and is said to give luster to the plumage in fitting birds for shows. Sunflower is greatly overrated for poultry purposes. It is an ungainly plant of no use for forage and its seed is so well liked by the sparrows that the only way to keep them till ripe is to cover the heads with netting.

Clipping Hens for Cleanliness.

My hens foul all the feathers below the vent; they appear healthy, but do not look nice. What can I do?

Take a pair of scissors and clip the fluff away from that part of the abdomen, give a teaspoonful of olive oil, and notice of they have any discharge that is of an offensive color or odor. Sometimes it is nothing but pure laziness with hens of the large breeds that causes this matting together of the fluff below the vent. We rarely see hens of the small breeds so affected. Whenever a hen soils her feathers clip her at once, and, in fact, it is a good custom to follow in any case. When hens are very heavily fluffed it interferes with the fertility of the eggs. In such cases there is not anything for it but the scissors.

Bowel Trouble in Chicks.

What is the cause of bowel trouble in young chicks, and what to do for it?

Bowel trouble in very young chicks is usually caused by a chill. It is very hard for us here to believe chicks get chilled because, not feeling the cold ourselves, we forget that chicks have really undergone a violent change from incubator to the outside atmosphere. In the Eastern States, great care is exercised in moving chicks from incubator to brooder oven, and also in seeing that the brooder itself is warm and fit to receive the chicks. But we are, as a rule, very careless in these little matters and the chicks feel the change and suffer from bowel trouble. Sometimes, of course, the trouble may be traced to the food, but more often it comes from a chill. The best way to cure it is to remove the chicks to new ground at once, or if in a brooder, clean it out well and spray with some disinfectant. Boil all the water that is given to the chicks and feed boiled rice once or twice a day in which a little cinnamon is mixed. Do not put in too much or they will not eat it, keep all meat away and just feed dry chick feed and boiled rice. No oatmeal or any other cereal but the rice; if chicks won't eat it, feed dry chick feed and boiled water and a little lettuce.

Quick Roosters and Laying Hens.

How can I get the young roosters off quick and the hens to lay in winter?

These two happy results come from correct methods of poultry keeping from the ground up. To get the cockerels off quick, they must be hatched from strong-germed eggs, incubated properly and kept growing from the first jump out of the shell. To get eggs in winter the pullets must come from the same conditions. Very few hens will lay in the early winter under any conditions. The pullets must be depended upon for that season and the hens kept properly will drop in some time in January.

Poultry Tonic.

What is a good poultry tonic?

The following is a very good tonic for general purposes: Tincture of red cinchona, 1 fluid ounce; tincture of chloride of iron, 1 fluid drachm; tincture of flux vomica, 4 fluid drachms; glycerine 2 ounces; water, 2 ounces. Mix and give one teaspoonful to a quart of water, allowing no other drink.

Poultry in the Orchard.

Kindly advise me about keeping hens in an orchard. I would like to know if they will injure the trees in any way if kept in large numbers. In what way would they benefit the trees?

From the point of view of the trees there is no doubt that they would be advantaged by the presence of the poultry, providing the coops are not allowed to interfere with the proper irrigation and cultivation. If it is practicable to handle the fowls in coops without causing the soil around the coops to become compacted by continual tramping, and if they are not kept upon the ground long enough to cause an excessive application of hen manure, which is very concentrated and stimulating, the result would unquestionably be beneficial. From the point of view of the tree, this benefit of injury would depend upon how long the fowls were kept around the tree and the maintenance of them in such a way that the soil should not become out of condition physically or too rich chemically for the satisfactory performance of the tree. If they can be moved frequently, and if they are only put in place when the soil is in such condition that tramping around the coops will not seriously compact it, the presence of fowls would be an advantage. On the other hand, if the coops are to be kept in place for a long time and all the ground outside of them crusted and hardened by tramping and the soil under the coops overloaded with droppings, the thrift and value of the trees will be seriously interfered with.

Caponizing.

Can three to four month old cockerels be caponized successfully in summer, and if so, what care, feed, etc., do they require afterwards?

The birds should be between two to three months, not over four, unless some very large variety that matures slowly. Size is equally important as age, and a bird to be caponized should not weigh more than one and a half pounds. The work can be successfully done in the summer season, but the fowl must be kept without food or drink for at least 24 hours, longer is better and keep in shady place. After caponizing, feed the bird what soft feed he will eat up and let him have plenty of water. Then leave him to himself as he will be his own doctor. In two or three days look them over and if there are any wind-balls, simply prick with a needle to let the air out; this may have to be done two or three times before the wound heals up, but after it has healed, treat just as you would other chickens and feed them about twice a day. There is nothing made by trying to rush nature; it takes fifteen months to grow a good capon of the large breeds.

Roup Treatment.

Up to a week ago the chickens had been exceptionally well in every way. Now they seem to have a cold and a running at the nose and with it a bad odor. It was suggested that this might be the beginning of roup, but I see no swell-head.

The distinguishing characteristic of roup is not so-called "swell head" or other form of cold, but the offensive roupy odor. When the cold has reached this stage it is a pronounced case of roup, and highly contagious. Separate all the ailing fowls and segregate them in comfortable hospital quarters, warm but with one side partly open for fresh air. Disinfect the quarters of the well fowls by spraying with distillate or cheap-grade coal oil and sprinkling the floors and about the houses with air-slaked lime. Use some simple remedy like coal oil or permanganate of potash to cleanse the throat and nostrils. With coal oil, first wipe the eyes and bill with a clean cloth dipped in the coal oil, then inject with a sewing-machine oil can enough coal oil to open and thoroughly clean out the nostrils. If the throat is affected, give a tablespoonful of sweet oil and coal oil, half and half, two or three times a day until relieved. One of our correspondents has sent us the following treatment with permanganate of potash which he has found the best roup remedy he has ever tried: Dissolve 1 ounce of permanganate of potash in 3 pints of water, hold the fowl's head in this for a second, then open the beak and rinse out the mouth in the solution. Wipe with a clean, soft cloth and apply a very little witch hazel or carbolated salve to the eyes, nostrils and head. Repeat the operation as often as the throat and head become clogged with mucus. Until the disease is eliminated from the premises, keep permanganate of potash in the drinking water of all the fowls, both sick and well. About 1 ounce to each 2 gallons of water or enough to give the water a claret color. The sick fowls should be allowed no other feed but a little stimulating mash three times a day. Where the fowls do not show a decided improvement in the course of a few days, or where the disease has assumed a violent form, all such birds should be killed and the bodies burned at once.

Bad Food for Chickens.

My chicks are about three weeks old and have always been strong and sturdy, but when taken sick first appear a little dumpish, then the head seems a little heavy and the neck lengthens out. As the disease advances they become staggery.

Your chicks have eaten soured food, decayed vegetables or tainted meat. Baby chicks are just like other babies and the same care should be used that their food be always sweet and fresh. Wet food should never be given chicks, nor raw meat nor anything the least bit tainted or stale. Put a teaspoon of coal oil in each pint of drinking water and see to it that the latter is kept pure and cool. Mix a teacup of sulphur with enough bran or shorts for each 100 chicks, moisten with sweet milk and feed it on clean boards, what the chicks will eat up clean in some, twenty minutes. Give them one feed of this each day for three days if the weather is dry. Clean the brooders and runs daily, then dust white with air-slacked lime and cover the lime with a sprinkling of clean sand. Rake and clean up the yards where they range and never let them eat any of their grain or food out of dirt and filth. You cannot doctor such small chicks and must depend upon the coal oil in the drinking water. Keep the water fresh, but add the coal oil until the chicks are relieved.

Open-Front Chicken Houses.

In what direction shall I face open-front poultry houses?

North or northeast is the proper direction to face the open fronts of poultry houses and coops in the Pacific Coast climate. The prevailing winds are from the south and southeast in the winter, and from the west and southwest in the summer. The occasional north winds or "northers," may be called dry winds, in fact, are an indication of dry weather, and so do not harm the fowls even when cold. We like the upper half of the north-end or slide of our poultry houses open with inch-mesh covering the open space and the eaves extending several inches as a protection. In case of an unusual storm from that direction, one thickness of burlap may be tacked to the edge of the extending eaves, and to the lower part of the opening. This will admit plenty of fresh air while breaking the force of the wind. We also have a large trap door for the use of the fowls, in the solid lower part of the open end, and the large door, for cleaning and sunning the house, in the west side.

A Point on Mating.

I have fine roosters a year old this April; would you advise keeping them for mating with the same hens next season, or do you advise selling each year and getting fresh stock?

The young males will be all right to mate with the same hens next season - that is, if they come through the molt with vigor. They will be just two years old and at their best. The molt is the test for both, hens and cocks. If they show no signs of ailing or weakness during that period, it is proof of the proper stamina and vigor.

Age for Mating.

At what age may a cockerel be mated with hens?

From nine months to a year is the proper age to mate a Leghorn cockerel.
Cockerels of the larger breeds should not be mated before a year old.

White-Yolk Eggs.

Why are eggs watery and light-colored?

The trouble is in the feed somewhere. Too much green feed, especially green feed that springs from wet, soggy ground, will sometimes make the eggs watery. Or if you are feeding more mash feed than dry grain, it will have that tendency. Some people claim that the feed a hen eats does not affect the egg at all; but if it does not, why do eggs differ in color and quality? Eggs that are laid by hens fed wholly on wheat, or the by-products of wheat, such as bran, shorts or middlings, all have a pale yolk. Now feed the hens some green feed - any kind will do - and the eggs from the same hens will have a yolk several degrees or shades darker.

Poultry Diarrhea.

Will you kindly tell me the cause and cure for bowel trouble among hens?

The "quick cure" for chick diarrhea has not yet been found. Prevention is the only sure remedy. The first treatment in diarrhea (which must not be confused with simple looseness of the bowels) should be a mild physic to clean out the digestive tract. Epsom salts is probably best for this purpose where a number of fowls are to be treated. This is usually given in the drinking water, but Dr. Morse, who has charge of the investigation of poultry diseases in the Bureau of Animal Industry, gives the following directions for administering the salts: "Clean out by giving epsom salts in an evening mash, estimating one-third to one-half teaspoonful to each adult bird, or a teaspoonful to each six half-grown chicks, carefully proportioning the amount of mash to the appetite of the birds, so that the whole will be eaten up quickly." For a few days afterward, feed only lightly with dry grain and tender greens, such as fresh-cut mustard and lettuce leaves. Keep plenty of pure, cool water, with just a thin skim of coal oil - one drop to each pint - for drinking; also plenty of sharp grit and fresh charcoal broken to the size of grains of wheat.

Limber-Neck.

A very peculiar disease is taking off my fowls. The head of the fowl bends down to the breast and the fowl looks like dead, there is also a slight discharge from the mouth. The head and tail droop and if the fowl could stand up they would almost touch.

When a fowl loses partial or entire control of the muscles of the neck the common name of the affection is limber-neck. In medical science limber-neck is regarded as a symptom rather than a disease, and may be due to a number of causes, such as derangement of the digestive organs, intestinal worms and ptomaine poisoning. The affected fowls should be given immediately a full tablespoon of fresh melted lard or sweet oil, to which has been added a scant teaspoonful, of coal oil. In an hour repeat the dose. For a few days the fowls should be fed on some light food, such as shorts scalded with sweet milk in which has been dissolved a level teaspoonful of baking soda to every pint of milk, and also allowed plenty of crisp, tender lettuce or similar greens. A little Epsom salts should be added to the drinking water for a few days. This treatment, if resorted to at the start, will be effectual, but if the poisoning has had its course long, nothing will save the bird.

Chicken Pox.

My one and two-year-old fowls are getting scabby combs. It starts with a round blackish spot and swells into many spots, finally nearly covering one side of the comb. Sometimes accompanying this is the closing of one eye, and later both eyes.

The trouble is chicken pox, which is a very contagious disease. A treatment which has been successful consists in bathing the sores with strong salt and water and giving the fowls a mash containing one teaspoonful of calcium sulphide for each 25 hens. With a large flock of hens the method successfully employed by one of the large coast ranches in stamping out an epidemic of the disease was to place a sulphur smudge, to which had been added a little carbolic acid, in the poultry house after the fowls had gone to roost. This was allowed to remain till the fowls began to sneeze, when it was instantly removed. The affected fowls were also treated by dipping the heads in a solution of permanganate of potash.

Roup in Turkeys.

My turkeys have a disease that is spreading rapidly. They commence with a running at the nose, have swelling under the eyes which are filled with pus.

This is clearly a case of cold developing into roup. Get one ounce of permanganate of potash and pour a quart of boiling water over; after it is cold, bottle for use. Now take an old tin can, three parts full of warm, not hot water, and drop in enough of the permanganate of potash to make it dark red. Hold the turk's head under in this can until it needs breath then give it time to breathe, and dip again. Press the fingers along the swollen parts towards the nostrils and get out all the pus you can, then take a sewing-machine oil can and fill it with a little of the mixture, and part olive oil, inject the liquid up the nostrils and in the cleft of the mouth. Put a little of the permanganate in the drinking water for all the flock. Make the water a light red, later it will turn to a dirty brown, but don't mind that.

Disinfectants.

What can I use to disinfect poultry belongings?

Sulphuric acid spray is good, but you will need to be very careful that you do not get it on the hands or clothing. Get 16 ounces sulphuric acid (50 per cent solution), water 6 gallons. Have the water in a wooden tub or barrel and add the sulphuric acid to the water very slowly, in order not to splash it on the flesh or clothes. But mind: nothing but wooden vessels to mix it in. When made according to directions, and of this strength it is a very valuable disinfectant, but is dangerous to use of any stronger mixing. After mixing, it can be stored in glass bottles or earthenware jugs. Another very good disinfectant for poultry houses and runs is the formaldehyde disinfectant. Formaldehyde 1 pint (40 per cent), water 2 gallons. This is fine for houses that you can shut up. Turn the fowls out of the building, close all windows, and spray thoroughly, then close the door and leave it do the work. Air well by opening windows and door several hours before the fowls go to roost.

Cloth for Brooding Houses.

Would some good grade of white cloth on a frame do as well, or would it be better than glass, for a brooder house, or would it keep out too much sun-heat?

Cheesecloth, not heavy cloth, would be better than glass, so far as the sun is concerned. There would be none of the overheating during the middle of the day followed by the chilling at night which are caused by a large expanse of glass. On the other hand, there should not be openings on opposite sides of the house to create a draft. Also, the rat and vermin question must be considered. It might be necessary to have wire screens made to fit firmly over the cloth at night.

Grains for Chickens.

What variety of grain adopted for poultry food will be the best to grow, with and also without irrigation?

Wheat is a standard grain for poultry feeding, and Egyptian corn is also largely used. Indian corn is also satisfactory, under the general roles for compounding poultry rations which are laid down by all authorities on the subject. Egyptian corn is very successful in the interior parts of the State, and, on lands which are winter-plowed and harrow to retain moisture, very satisfactory results can be secured by summer growth without irrigation from planting as soon as frost danger is over.

Plucking Ducks and Geese.

I would like to know about how, when and how often to pick old ducks so as to get the feathers for pillows and not kill the ducks, either. Will they lay any eggs while growing new feathers?

Neither ducks nor geese should be plucked until after the laying season is over, which will be in July. Just before the moult, when the feathers begin to loosen, they may be plucked again. Those most considerate of their birds make only this latter plucking, which does not greatly inconvenience the fowls. At no time must they be plucked unless the feathers are "ripe"; that is, dry at the root, so that no bleeding or injury to the skin is caused. An old stocking is drawn over the head of the victim, and the bird held in the plucker's lap on a burlap apron; then the soft feathers on the body are quickly and very gently removed; but those on the side of the body which support the wings should not be taken. Great care should be exercised not to injure the skin or pinfeathers or pull the down. To grow new feathers quickly and resume laying are matters which depend largely upon the condition of the bird and the feed. The latter should consist of some 15 per cent of animal food.

Feeding Hens for Hatching Eggs.

Should soft feed be given to the mothers of chicks intended for broilers? How about dry mash? How would you advise feeding animal protein?

Cut out all ground feed, except perhaps a little wheat bran. While you may not get quite as many eggs, they will all have good strong germs and the chicks will stand forcing to the limit, while if you force the egg output you reduce the vitality of the germs and livability of chicks hatched. The only way to feed hens whose eggs are intended for hatching chicks for broilers is to feed whole grain and make them exercise for it, good green feed, or, better still, sprouted oats, and feed beef scrap in a hopper all the time. At first, while it is new, they may eat more than you would give them but don't mind that they will regulate the quantity in a few days better than you can. Get a good grade of beef scrap and keep it in a hopper that will not let rain in or keep it under cover and feed all the wheat and oats they require; if you are short on green feed give them a bale of alfalfa hay to work on.

A Dry Mash.

Will you give a formula for a dry mash?

Wheat bran, 500 pounds; middlings, 200 pounds; cracked corn, 200 pounds; charcoal, 20 pounds; alfalfa meal 200 pounds; bone meal, 150 pounds; blood-meal 100 pounds; meat cracklings, if ground, 200 pounds; ground oats or barley, 300 pounds. Give oyster shell separately and supply fowls with good sharp grit.

Depluming Mites.

My chickens are losing the feathers from their necks, some three inches down the front and then extending around the neck.

The loss of feathers is probably due to the depluming mite. Dust well with buhach through the feathered portion of the bird and apply carbolated vaseline to the bare skin and the edges of the feathers where the insects work. Do this daily as long as needed. When vaseline is not on hand, a mixture of coal oil and sweet oil applied with a soft sponge squeezed nearly dry does as well. We would advise that you make a general cleaning and spraying of your poultry quarters, nest boxes, etc.

Part IX. Pests and Diseases of Plants

Control of Grasshoppers.

This county is having trouble with the grasshoppers as are other counties. Would you kindly inform me what I could do to exterminate them on my young orchard?

The best thing for grasshoppers is to fix up a lot of poison. This is made in the proportion of 40 pounds of bran, 2 pounds of molasses and 5 of arsenic, mixed together as a mash. They will take this wherever they find it, even when nice green leaves are close by, but it has to be kept moist. Grasshoppers can also be reduced by driving a "hopper doser" over ground where they are. This is made somewhat like a Fresno scraper, but is much longer and the bottom is covered with crude oil. When disturbed the hoppers jump up and fall into the oil. Besides the poison, you should also protect the trunk of the tree to prevent the hoppers from climbing up it. This can be done by applying tree tanglefoot, or putting on one of the tree guards that prevent climbing insects from passing up to the leaves. The combination of poison and tree guards will give you about all the protection you need.

Sunburn and Borers.

Please state the best remedy for keeping the borer out of young fruit trees.

Sunburn can be prevented in many ways. The manufactured tree-protectors are good if they are light colored and are kept in place so that the sun does not scald above or below them. Wrapping spirally with narrow strips of burlap, torn from old grain sacks, from the base to the forking of the branches, is also good. A very effective and widely used method is to apply a good durable whitewash which may be made of 30 pounds of lime, 4 pounds of tallow and 5 pounds of salt, adding the salt to the water used in slaking the lime, stirring in the tallow while the slaking is in progress and hot, and then adding water to thin the wash so that it will work well with pump or brush.

Gumming of Prune Trees.

I write to ask for information concerning my prune trees. They are from two to six years old and the gum is exuding from them. As I notice the branches dying I cut them out, but this doesn't seem to save the tree. I would appreciate any information you can give me.

This is a pretty hard matter to diagnose from a distance. There is a good probability that the trouble is caused by sunburn, a point you could determine on inspection. Whitewash would be a protection against this and more or less of a cure also. Furthermore, borers may be the cause, which can be determined by examining the points where the gum exudes, seeing if any wood grains are present. These borers should be dug out and whitewash applied, which latter also protects against this trouble. Lastly, your ground may be drying out, which also you can determine and remedy.

Borers in Olive Twigs.

There are quite a number of olive trees in this locality that have something wrong with them. They make a growth of five or six inches and the center twig dies back, then it sprouts out at the sides and makes another growth in the same way. This makes a thick bush instead of the tree coming up as it should.

The dying back is caused by a beetle which bores into the twigs. The twigs above the point where the beetle enters dies and then, of course, buds come out from healthy wood below. No treatment has been devised against it, though its breeding ground is limited if all dead wood and brush and litter is cleaned up and twigs are cut off below the point of injury whenever the work of the insect is seen.

Raspberry Cane Borer.

Can you tell me what to do for my Loganberries and raspberries? A small worm got into them in the new growth of wood lost summer, right in the tips of the new growth of wood, and then worked down through the pith of the wood, and as fast as they worked down the can wilted.

This is the raspberry horn-tail, or the cane-borer. The adults are wasp-like insects about a half-inch long and very active. They come out of the canes in spring and the females soon lay eggs in the tender tips of the young shoots. These eggs soon hatch and the larvae eat their way up toward the tip, which causes it to wither and die. It is this injury that causes much notice. As the tip dies, the larvae turn and go down into the canes, as in the sample sent, also injuring them greatly, though possibly not killing them for some time. The only way to attack them is to pinch the spots where the eggs were laid; then those that escape and cause the tips to wilt should be destroyed by cutting off the tips below the point of injury or cutting off the canes when they show damage. Likewise, the insects work on the wild rose, and cutting all those out around a place will prevent enough adults from developing to permit little damage to be done, always provided the berries are well looked after.

Control of Red Spider.

Can you give directions for the prevention of injury by the red spider to almond and other trees in the Sacramento volley?

The red spider on almond and prune trees is usually controlled by the thorough application of dry sulphur to the foliage. On almonds the first sulphuring should be done as soon as the leaves appear in March. A second application is advised from the 1st to the 10th of May. A third application should be made from the 1st to the 10th of June. Prune trees should be treated as soon as the spider appears. In the Sacramento valley this usually occurs about the first week of July. Full-grown trees require about a pound of sulphur which should be thoroughly distributed throughout the foliage. The old method of throwing a handful of sulphur in the branches of the tree or on the ground under the tree is valueless. The use of a blower is economical in large orchards, but a can with perforated bottom is frequently used on young trees or small orchards with good results. In normal seasons the spider is easily, controlled by dry sulphuring. When the pest does not yield to this treatment, a spray is recommended.

Liquid Spray for Red Spider.

Is there any liquid spray I can use in my spraying that will kill the red spider without injuring the foliage of the almond?

A liquid spray for red spider is made by taking sulphur 30 pounds; lime (reduced to milk form by water), 15 pounds; water, 200 gallons; or use commercial lime-sulphur, 4 or 5 gallons to 200 gallons of water. These sprays can be applied without injuring the foliage. They are more expensive in labor cost than dry sulphuring, but are more effective.

Apple-Leaf Aphis.

I am sending herewith a small piece from one of my young apple trees. If you can, will you kindly tell me what the insects are an it, and what I had better do for them?

The apple twig which you send is infested with the eggs of the leaf aphis or leaf louse. These eggs are very difficult to kill. A good thorough spraying with lime-sulphur might, however, get rid of many of them and would be good for the trees otherwise - diluting according to condition of tree growth. The chief campaign against the leaf aphis, however, must be made early in the growing season, just as these pests are beginning to hatch out and to accumulate under the leaves of the new growth. They should then be attacked with properly made kerosene emulsion or tobacco extract with a nozzle suited to land the spray on the under side of the leaves. Unless these pests are attacked early in the season and repeated if necessary, your apples on bearing trees will be ruined so far as they attack them, being small, misshaped and worthless. On young trees the destruction of the foliage is fatal to good growth.

Woolly Aphis.

Will you kindly inform me what you consider the best treatment for apple trees affected by woolly aphis?

The best way to kill the woolly aphis on the roots is to remove the earth from around the tree to a distance of one or two feet, according to the size of the tree, digging away a few inches of the surface soil, Then soak the soil around the tree with kerosene emulsion, properly made, of 15 per cent strength, and replace the earth. Be sure you get a good emulsion, for free oil is dangerous. For the insects above ground on the twigs, a good spraying while the tree is out of leaf will kill many, but some will survive for summer spraying, and for this a tobacco spray may be most convenient.

Blister Mite on Walnuts.

I am sending you some walnut leaves with some swellings an them. They are very plentiful on some trees here. Is the trouble serious and will it spread?

This is merely Erinose, or Blister Mite, which is a very common trouble on walnuts, but does not do enough damage to call for methods of control. These swellings are caused by numerous, very small insects which live within the blisters on the under side of the leaf amongst a felt-like, heavy growth which develops there. While this effect is very common, it produces no appreciable injury and needs no treatment for its control.

Scale on Apricots.

I would like to know how to check the scale on apricot trees.

The most common scale on apricots, the brown apricot scale, is usually held in check by the comys fusca, which is as widely distributed as the scale itself. If it gets beyond the parasite, you should spray in winter with crude oil emulsion. If some scales are punctured or have a black spot on top, the comys fusca is busy and you probably will be safe enough without doing anything.

Fumigating for Black Scale.

I would like to know the best method of eradicating the black scale from my orange trees, whether by spraying or fumigation?

Spraying has been given up as a suitable method for controlling the black scale on citrus trees, and the only recognized method of merit where the scale is bad is by fumigation with hydrocyanic acid gas. You should communicate with your county horticultural commissioner, who, through inspectors, will see that you have a good job done, at the right time and at as moderate price as is compatible with good work. It is impossible to 'eradicate' the black scale, but there is a great difference in the amount that can be killed, and it pays to have a job done as near perfectly as possible. Similar methods of attacking other scale insects on citrus trees are used.

Finding Thrips.

How can the presence of pear thrips be detected in a prune orchard? Will the distillate emulsion-nicotine spray control brown scale as well as thrips?

You can find thrips by shaking a cluster of blossoms, as soon as they open, over a sheet of paper or in the palm of your hand. The thrips are very minute, transparent, somewhat louse-like insects. The spray you mention would probably have little effect on the brown scale which would still be in the egg state and under cover, at the time the early spring spraying for the thrips.

Control of Pear Slug.

I am sending, under separate cover, some samples of cherry tree leaves that have been attacked by a small snail or slug. Kindly let me know what they are, and how to rid the trees of them.

The creatures you speak of are the pear slugs, or the cherry slugs, as they are sometimes known. Although slimy, like the big yellow slug that is a pest in vegetable gardens, it is no relation thereto, but is the larva of an insect. Its olive green color, slimy appearance and the way it eats the surface of the leaves make it about the easiest of all insects to identify. Parasites and predacious insects usually keep it in fair control. Whenever artificial methods of control are needed the slugs can best be destroyed by sprinkling dust of any kind upon them. If you can get a machine for sulphuring a vineyard and use some air slaked lime or other fine dust, it will fix them quickly and inexpensively, though any way of applying dust may be used.

Cutworms and Young Trees.

What method should be used to protect young fruit trees from cutworms?

Hoe around the trees or vines and kill the fat, greasy grubs which you will find near the foliage. Put out a poisoned bait which the worms like better than the foliage, viz. Bran, 10 pounds; white arsenic, 1/2 pound; molasses, 1/2 gallon; water, 2 gallons. Mix the arsenic with the bran dry. Add the molasses to the water and mix into the bran, making a moist paste. Put a tablespoonful near the base of the tree or vine and lock up the chickens.

Control of Squash Bugs.

We are troubled with pumpkin bugs. Please tell us what to do for them.

When the bugs first make their appearance in the field they can be easily disposed of by hand picking and dropping into a bucket containing about two inches of water with about one-fourth inch of kerosene on top to kill the bugs. The picking should be done in the morning, as the bugs are apt to fly in the warm part of the day and scatter where already picked. Two persons can pick over an acre in one and a half hours, and two pickings are usually sufficient for a season, as after the vines begin to run over the ground pretty well the bugs will not be able to hurt them much. A pair of thin old gloves will help to keep off one's hands some of the perfume from the bugs. The sooner the work starts the fewer bugs to pick. Cleaning up of all old vines in the fall and removing litter in which the mature bugs hide for the winter will permit less eggs to be laid in the spring and there will be fewer bugs to pick as a result.

The Corn Worm.

Last year all my ears of corn were infested with maggot, growing fat thereon. Can you help me scare them away?

You have to do with the so-called corn worm which is very abundant in this State and one of the greatest pests to corn growing. It is the same insect which is known as the boll worm of the cotton in the Southern States. No satisfactory method of controlling this has been found, although a great deal of experimentation has been done. Nearly everything that could be thought of has been tried without very satisfactory results. A late planted corn has sometimes been free, for the insect is not in the laying stage then. If it were not for this insect the canning of corn would be an important industry in this State.

Melon Lice.

I have in about four acres of watermelons, and there seem to be lice and a small gnat or fly, and also some small green bugs and white worms on the under part of the leaves, which seem to be stopping the growth of the vines, making them wilt and die. They seem to be more in patches, although a few on all the vines. Can you please tell me what to do for them?

Melon lice are very hard to catch up with after you have let them get a start. Spraying with oil emulsions, tobacco extracts, soap solutions, etc., will all kill the lice if you get it onto them with a good spray pump and suitable nozzles for reaching the under sides of the leaves. The gnats you speak of are the winged forms of the lice; the white worms may be eating the lice; the "small green bugs" may be diabroticas. If you had started in lively as soon as you saw the first lice you could have destroyed them in the places where they started. Now your chance lies largely in the natural multiplication of ladybirds and the occurrence of hot winds which will burn up the lice. It is too late probably, to undertake spraying the whole field.

Wire Worms.

Is there any way to destroy or overcome the destructive work of the wireworm, which I find in some spots takes the lion's share of crops, such as beans, potatoes, onions, etc.?

We do not know any easy way with wire worms. Nitrate of soda is believed to kill or repel them, but you have to be careful with it, for too much will either over-stimulate or kill the kill; about 200 pounds per acre, well distributed, is the usual prescription for the good of the plants. Wire worms can probably be killed with carbon bisulphide, using a tablespoonful poured into holes about a foot deep, three or four feet apart. The vapor would permeate the soil and kill all ground insects, but the acre-cost of such treatment must be measured in its relation to the value of the crop. The most promising policy with wire worms is rotation of crops, starving them out with a grain or grass crop and not growing such crops as you mention continually on the same land.

Bean Weevil.

How can I keep certain insects from getting into my dry beans? I have finished picking the crop. Every year a little, short, stubby beetle gets in them before spring and makes them unfit for use.

You have to do with the bean weevil. The eggs are inserted by the insect while the beans are still green in the pods; subsequently the eggs hatch and the worm excavates the interior of the ripened beans. The beans can be protected after ripening by heating carefully to 130° Fahrenheit, which will destroy the egg, or the larva if already hatched. Of course, this heating must be done cautiously and with the aid of a good thermometer for fear of destroying the germinating power. The work of the insect can also be stopped by putting the beans in a barrel or other close receptacle, with a saucer containing about an ounce of carbon bi-sulfid to vaporize. Be careful not to approach the vapor with a light. After treatment for one-half hour, the cover can be removed and the vapor will entirely dissipate. This is a safer treatment than the heating. Similar methods of control can be used on other pea and bean weevils.

Slugs in Garden.

Can you advise me how I can get rid of slugs in my garden?

When barriers of lime, ashes, etc., are ineffective, traps consisting of pieces of board sacking and similar materials placed about the field prove inviting to the slugs. They collect under these and by going over the field in the early morning they may be put into a salt-water solution or otherwise destroyed. Arsenical sprays applied with an underspray nozzle to the lower surface of the leaves will help control the slugs. Poison bran mash consisting of 16 pounds of coarse bran, 2 quarts of cheap syrup, and enough warm water to make a coarse mash, is very good for cutworms and should be equally effective for slugs. It should be placed in small heaps about the plants to be protected. Cabbage leaves dipped in grease drippings and placed about the fields also prove attractive bait for the slugs, which may then be collected there. If a person has a taste for poultry, the keeping of a few ducks may solve the slug problem without further bother. Cultivation or irrigation methods that give a dry surface most of the time also discourage these pests.

Cause of Mottle Leaf.

What is the cause and cure of mottle leaf of citrus trees?

There are apparently a number of causes of this trouble, all more or less obscure and hard to overcome. It is generally thought that it is due to poor nutrition, whatever the reason for poor nutrition might be. The presence of a nematode or eel worm on the roots has found to be a cause of mottle leaf in many cases. Poor drainage, too sandy soil and a number of other things frequently cause it. Whatever the cause, no one good method of cure has been found.

Potato Scab.

I think most of my potatoes will have some scab. Will you please tell me if my next crop would be apt to have scab, provided I got good clean seed and planted in the same ground?

It seems demonstrated that a treatment of the seed will practically insure against potato scab. One method is dipping the potatoes in a solution of corrosive sublimate. Dissolve one ounce in eight gallons of water and soak the seed potatoes in this solution for one and one-half hours before cutting.

Gopher Poison.

I have some alfalfa, some hogs and some gophers, also some strychnine and carrots. If I put the strychnine on the carrots, and endeavor to poison the gophers, and the hogs get hold of the poison will it kill them?

You will find that hogs are liable to poison like any other animal, and the safest way to poison the gophers, while the hogs are running in the field is to bury the poisoned carrots very deeply in the gopher hole and then put a row of sticks or branches over the mouth of the hole so that the hogs cannot root around and get at the poisoned carrots.

How to Make Bordeaux.

Use copper sulphate (bluestone) 5 pounds; quick-lime (good stone lime), 6 pounds; water, 50 gallons. Put the bluestone in a sack and hang it so it will be suspended just under the surface of a barrel of water over night, or dissolve in hot water. Use one gallon of water to one pound of bluestone. Slake the lime in a separate barrel, using just enough water to make a smooth, clean, thin whitewash. Stir this vigorously. Use wooden vessels only. Fill the spray tank half full of water, add one gallon of bluestone solution for each pound required, then strain in the lime and the remainder of the water and stir thoroughly. The formula may be varied according to conditions, using from 3 to 8 pounds of bluestone to 50 gallons of water and an equal or slight excess of lime. Use the stronger mixture in rainy weather. Keep the mixture constantly agitated while applying.

Formula for Lime-Sulphur.

To make lime-sulphur take quick-lime, 20 pounds; ground sulphur, 15 pounds and water 30 gallons. Slake the lime with hot water in a large kettle, add the sulphur and stir well together. After the violent slaking subsides add more water and boil the mixture over a fire for at least one hour. After boiling sufficiently strain into the spray tank and dilute with water to the proper strength. If a steam boiler is available, this mixture may be prepared more easily on a large scale by cooking in barrels into which steam pipes are introduced. This mixture cannot be applied safely except during the winter when the trees are dormant. A large proportion of the lime-sulphur used in the State is purchased already prepared in more concentrated form.

Index

Fruit Growing.

Almond
    Grafting on Peach
    Pruning

in the Hills
Increase Bearing
Temperature and Moisture
from Seed
High-grafted

Vegetable Growing.

Artichokes
   Jerusalem
 Globe
 Growing
Asparagus Growing
Beets
   Leases for Sugar
 Topping Mangel Wurzels
Brussels Sprouts - Blooming
Bean
   Growing
   Hoeing
 as Nitrogen Gatherer
 Yard-Long
 Why Waiting
 Blackeye
 Are Cow-Peas
 Horse-Bean Growing
 Growing Castor
 Inoculation
 On Irrigated Mesas
California Grown Seed
Cloth for Hotbeds
Celery, Blanching
Chili Peppers
Corn

Soils, Fertilizing and Irrigation.

and Oranges
  Crop Changes
Moisture Defects
Refractory
Suitable for Fruits
Blowing
Improving Heavy
Reclaimed Swamp
  Improving Uncovered
Sand for Clay
Sour
and Old Plaster
Handling Orchard
Depth for Citrus
Summer Fallow
Sub-soil, Plow for
Stable Drainage for Fruit
Seeds, Soaking
Trees
  over High-water
Plowing toward or from
Irrigated or not
Too Much Water
Too Little Water
Thomas Phosphate, Applying
Water
  Artesian
from Wells or Streams

Live Stock and Dairy.

Buttermilk Paint
Butter
  Going White

Feeding Farm Animals

Depraved Appetite
Dentist Needed
Dehorning
Forage Poisoning
Fungus Poisoning
Fly Repellants
Flea Destroyers
Garget
Gland Enlarged
Heaves
Horse with Itch
Horses Feet, Treatment
Hog Cholera
Hog Sickness
Infectious Mastitis
Irritation of Udder
Injury to Udder
Kidney Trouble
Lumpy jaw
Lumps in Teat
Loss of Cud
Mange, Is it
Mangy Cow
Musty Corn for Pigs
Nail Puncture
Neck Swelling
Pregnancy of Mare
Paralysis
Pneumonia in Pigs
Paralysis of Sow
Rickets in Hogs
Scabby Swelling
Skin Disease, Fatal
Scours
Side-bone
Shoulder injury
Stiff joints
Swelling in Dewlap
Sterile Cow

Raspberry Cane Borer
Sunburn and Borers
Scale on Apricots
Spray for Red Spider
Slugs in Garden
Thrips, Finding
Wooly Aphis
Wire Worms